D1711097

Issues in Higher Education

Series Editor: GUY NEAVE, International Association of Universities, Paris, France

Other titles in the series include

GOEDEGEBUURE et al.
Higher Education Policy: An International Comparative Perspective

NEAVE and VAN VUGHT
Government and Higher Education Relationships Across Three Continents: The Winds of Change

SALMI and VERSPOOR
Revitalizing Higher Education

YEE
East Asian Higher Education: Traditions and Transformations

DILL and SPORN
Emerging Patterns of Social Demand and University Reform: Through a Glass Darkly

MEEK et al.
The Mockers and Mocked? Comparative Perspectives on Differentiation, Convergence and Diversity in Higher Education

BENNICH-BJORKMAN
Organizing Innovative Research? The Inner Life of University Departments

HUISMAN et al.
Higher Education and the Nation State: The International Dimension of Higher Education

CLARK
Creating Entrepreneurial Universities: Organizational Pathways of Transformation.

GURI-ROSENBLIT
Distance and Campus Universities: Tensions and Interactions. A Comparative Study of Five Countries

TEICHLER and SADLAK
Higher Education Research: Its Relationship to Policy and Practice

TEASDALE and MA RHEA
Local Knowledge and Wisdom in Higher Education

TSCHANG and DELLA SENTA
Access to Knowledge: New Information Technology and the Emergence of the Virtual University

TOMUSK
Open World and Closed Societies: Essays on Higher Education Policies "in Transition"

HIRSCH and WEBER
Challenges facing Higher Education at the Millennium

The IAU

The International Association of Universities (IAU), founded in 1950, is a worldwide organization with member institutions in over 120 countries. It cooperates with a vast network of international, regional and national bodies. Its permanent Secretariat, the International Universities Bureau, is located at UNESCO, Paris, and provides a wide variety of services to Member Institutions and to the international higher education community at large.

Activities and Services

- IAU-UNESCO Information Centre on Higher Education
- International Information Networks
- Meetings and seminars
- Research and studies
- Promotion of academic mobility and cooperation
- Credential evaluation
- Consultancy
- Exchange of publications and materials

Publications

- International Handbook of Universities
- World List of Universities
- Issues in Higher Education (monographs)
- Higher Education Policy (quarterly)
- IAU Bulletin (bimonthly)

Higher Education, Research, and Knowledge in the Asia Pacific Region

Edited by

V. Lynn Meek
and
Charas Suwanwela

HIGHER EDUCATION, RESEARCH, AND KNOWLEDGE IN THE ASIA PACIFIC REGION
© V. Lynn Meek and Charas Suwanwela, 2006.

All rights reserved. No part of this book may be used or reproduced in any manner whatsoever without written permission except in the case of brief quotations embodied in critical articles or reviews.

First published in 2006 by
PALGRAVE MACMILLAN™
175 Fifth Avenue, New York, N.Y. 10010 and
Houndmills, Basingstoke, Hampshire, England RG21 6XS
Companies and representatives throughout the world.

PALGRAVE MACMILLAN is the global academic imprint of the Palgrave Macmillan division of St. Martin's Press, LLC and of Palgrave Macmillan Ltd. Macmillan® is a registered trademark in the United States, United Kingdom and other countries. Palgrave is a registered trademark in the European Union and other countries.

ISBN-13: 978–1–4039–7095–4
ISBN-10: 1–4039–7095–5

Library of Congress Cataloging-in-Publication Data

    Higher education, research, and knowledge in the Asia Pacific region /edited by V. Lynn Meek and Charas Suwanwela.
      p. cm.—(Issues in higher education.)
    Includes bibliographical references and index.
    ISBN 1–4039–7095–5 (alk. paper)
      1. Education, Higher—Asia—Research. 2. Education, Higher—Pacific Area—Research. 3. Higher education and state—Asia. 4. Higher education and state—Pacific Area. I. Meek, V. Lynn (Vincent Lynn). II. Suwanwela, Charas.

LA1058.H54 2006
378.5—dc22   2006045595

A catalogue record for this book is available from the British Library.

Design by Newgen Imaging Systems (P) Ltd., Chennai, India.

First edition: December 2006

10 9 8 7 6 5 4 3 2 1

Printed in the United States of America.

Contents

List of Tables — ix
List of Figures — x
Foreword — xi
List of Contributors — xiii

1. Introduction — 1
   V. Lynn Meek and Charas Suwanwela

2. Modernization, Development Strategies, and Knowledge Production in the Asia Pacific Region — 27
   William K. Cummings

3. Research Policy and the Changing Role of the State in the Asia Pacific Region — 43
   Grant Harman

4. The Changing Landscape of Higher Education Research Policy in Australia — 65
   V. Lynn Meek

5. Policy Debate on Research in Universities in China — 91
   Wei Yu

6. Between the Public and the Private: Indian Academics in Transition — 107
   Karuna Chanana

7. Development and Impact of State Policies On Higher Education Research in Indonesia — 135
   Jajah Koswara and Muhammad Kamil Tadjudin

8. National Research Policy and Higher Education Reforms in Japan — 153
   Akira Arimoto

9. Acknowledging Indigenous Knowledge Systems in
   Higher Education in the Pacific Island Region   175
   Konai Helu Thaman

10. Higher Education Research in the Philippines: Policies,
    Practices, and Problems   185
    Rose Marie Salazar-Clemeña

11. Higher Education Reform in Thailand   201
    Charas Suwanwela

12. Conclusion: Research Management in the Postindustrial
    Era: Trends and Issues for Further Investigation   213
    V. Lynn Meek

Index   235

List of Tables

2.1	Distribution of government R&D budget appropriations in selected countries by socioeconomic objective: 2000 or 2001	33
2.2	Science and engineering articles, by region and country/economy: 1988–2001	39
3.1	Size of selected Asia and Pacific higher education systems as measured by total enrollments, teaching staff, and graduates	46
3.2	Researchers and R&D expenditure in selected Asia and Pacific nations	47
3.3	Summary results from OECD survey on patenting and licensing activities in universities and public research institutes	60
4.1	Overview of international R&D performance by OECD countries, 2000	69
4.2	GERD performed by sector and OECD country, 2000	70
4.3	Researchers per 10,000 labor force by sector of employment	70
4.4	University expenditure on research and experimental development by source of funds, 1988–2000 (percentages)	71
4.5	Research income by source, 2001 (percentages)	71
4.6	University revenue by source 1995–2000 (AU$ in billions) (adjusted by CPI to 2000 terms)	75
4.7	Higher education institution operating revenue by source, 2003	75
4.8	University expenditure on research and experimental development by type of research activity, 1988–2000 (annual expenditures are in respective year prices)	87
4.9	Socioeconomic objective of research by type of funds (percentages)	87
7.1	Relevance of current research and development programs	150
8.1	Selected countries' shares of published papers (percentages)	167
8.2	Selected countries' shares of citations (percentages)	167
11.1	The budget for research outlined in the 7th National Development Plan	206

List of Figures

5.1	Change in the gross enrollment ratio (GER) and the total number of students (1990–2004)	93
5.2	Increase of R&D expenditure in universities (1985–2002) (billion Yuan)	97
7.1	One-year research type 1988–2003	142
7.2	Mulit-year research type 1991–2003	143
7.3	Applications of one-year type of appropriate technology (Voucher) 1994–2003	147

Foreword

The Asia Pacific region is experiencing a period of rapid and far-reaching economic and social change, driven by an emerging knowledge-based society, accelerating globalization, more market-oriented systems (including higher education), and increasing international economic competition. The emerging knowledge-based society enhances the value of knowledge through encouraging its discovery, dissemination, application, and understanding. Impressive developments are taking place in the application of new information and communication technology. Globalization is bringing about increasing competition among all nations with regard to science, technology, culture, and education, while at the same time necessitating mutual understanding and cooperation. Market-oriented approaches through defining social relationships in terms of supply and demand mechanism are highlighting the importance of economic rationalization, efficiency, relevance, and accountability for higher education institutions and systems. These and related issues are what shape if not determine the character, structure, and direction of higher education, research and knowledge. Over the past 2 years and a bit more, a committee of the UNESCO Forum on Higher Education, Research and Knowledge has been attempting to understand and analyze the impact of these issues on the higher education systems of the Asia Pacific.

The objective of the UNESCO Forum on Higher Education, Research and Knowledge, which was initiated in 2003, is as follows: "To widen the understandings of systems, structures, policies, trends and developments in higher education, research, and knowledge, particularly in developing countries." Within these areas, the work of the Forum is focused on gathering and engaging with existing and ongoing research; identifying research gaps and new priorities; stimulating and facilitating research; bringing to the fore current issues and debates; making available research findings; and disseminating information on policies and practices.

The Forum provides a means for all participating countries to analyze similarities, differences, strengths, and weaknesses of their respective countries' higher education, research, and knowledge systems on a regional and global basis. The are five regional committees that meet regularly and exchange information and understanding through such mechanisms as seminars, colloquiums, commissioned research, and publications. A global perspective is maintained through the regional committee chairs participating in the annual Global Scientific Committee at the UNESCO headquarters in Paris.

The Regional Scientific Committee for Asia and the Pacific began its activities in 2003 with the following objective: "To function as an intellectual platform, with some of the central aims being: conceptualizing research areas or topics; engaging in public and institutional debate and discourse; conceptualizing short and long term strategies to highlight research and concerns within the project framework." The first Committee meeting took place in Bangkok in February 2003, followed by the second meeting in New Daly later that year, and the fourth meeting in Seoul in 2005. The third meeting and first Committee sponsored research seminar was held at the United Nations University, in Tokyo, in May 2004. Through these organized and intensive activities Committee members have analyzed a variety of the higher education research and knowledge production problems their respective countries face, individually and collectively. The various Committee meetings have articulated a detailed research agenda guiding the Committee's efforts. This book is based on the efforts to date of the Committee members at understanding and analyzing the higher education, research and knowledge issues of the Asia Pacific region and includes the intellectual contribution and reflections of invited participants in the Tokyo seminar as well. All of this is explained in much more detail in the Introduction to this volume.

Committee members have gained much from their interactions with one another and have greatly extended their knowledge of the higher education research and knowledge issues facing the Asia Pacific region. In the spirit of the Forum, the Committee now feels obliged to share its understandings and the knowledge gained so far with others in the region and Globally. It is for these reasons that Committee members have prepared this publication.

As Chair of the Regional Scientific Committee for Asia and the Pacific I would like to express my heartfelt thanks to the two Committee members who undertook the editorial responsibilities for preparing this book, Professor Lynn Meek and Professor Charas Suwanwela. On behalf of the Committee, I wish to thank as well Dr Katri Pohjolainen Yap, former Senior Program Specialist and Project Manager for the UNESCO Forum on Higher Education, Research and Knowledge, for her very hard work and wonderful leadership. Through me the Committee also expresses its appreciation for the services of Terry Tae Kyung Kim, former Assistant Program Specialist, who died far too young. This book would not have been realized without these two persons' contribution to the Committee's work and it is to them that the book is dedicated.

Finally, and on behalf of the Regional Scientific Committee for Asia and the Pacific, I am happy to have this opportunity of expressing my sincere gratitude to all the contributors for their excellent papers. The book will provide a source of significant and creative material on higher education, research and knowledge for a variety of audiences interested in the Asia Pacific.

AKIRA ARIMOTO

LIST OF CONTRIBUTORS

Akira Arimoto is Professor and Director of the Research Institute for Higher Education at Hiroshima University, Japan. He is President of the Japan Association of Educational Sociology; President of Japanese Association of Research Institute for Higher Education; former President of Japanese Association of Higher Education Research; Steering Committee Member of Eight-Nation Education Research Project. He specializes in the sociology of education and sociology of science, and teaches Comparative Higher Education in the Graduate School of Education, Hiroshima University. He has published many books and articles. He is the Chair of the UNESCO Scientific Committee for Asia and the Pacific.

Karuna Chanana is Professor, Sociology of Education, Zakir Husain Center for Educational Studies, School of Social Sciences, Jawaharlal Nehru University, New Delhi, India. She has published in the areas of the sociology of education; equity and diversity; educational policy and process with a focus on societal change and affirmative action; links between higher education and basic education; international flow of students in India; women's higher education: recruitment and relevance; women and research; women's work, education and family strategies in the context of change and mobility; sexual harassment of university campus; women in the police force: gender positive initiatives in Delhi Police. She is a member, Research Advisory Committee, State Council for Educational Research and Training; former member, Programme Advisory Committee of the National Council of Educational Research and Training; member, editorial advisory boards: Sociological Bulletin, Social Change, and Journal of Educational Planning and Administration. She is a member of the UNESCO Scientific Committee for Asia and the Pacific.

Rose Marie Salazar-Clemeña is the Executive Vice-President, De La Salle-College of Saint Benilde, The Philippines. She is a University Fellow and Full Professor of De La Salle University-Manila. She obtained her PhD in Psychology from the University of Minnesota. She has published in areas of teacher education, teaching and learning in higher education, counseling psychology, counselor education, health psychology, career psychology, psychometrics, program evaluation, educational management, peace and values education, and research in higher education. She has served as President of the Association of Psychological and Educational Counsellors of Asia-Pacific and other professional organizations and has been a consultant for projects of the UNESCO as well as the Commission on Higher Education and the Department of Education of the Philippines. She is Vice-Chair of the UNESCO Scientific Committee for Asia and the Pacific.

William K. Cummings is Professor of International Education and International Affairs, George Washington University. He received his PhD from Harvard University with a dissertation on "The Academic Marketplace and University Reform in Japan" in 1972. Dr. Cummings has been involved in development work for over 25 years, including long-term residence in Ethiopia, India, Indonesia, Japan, and Singapore, and short-term consultancies in over 15 countries in Asia, the Middle-East, Africa, Latin America, and Eastern Europe. Included in this work is experience in developing higher education programs and monitoring their progress for OECD, the World Bank, USAID, and the Ford Foundation. He has authored or edited over 100 articles and 22 books or monographs on education and development including *The Institutions of Education* published by Symposium Books in 2004. Dr. Cummings is past President of the Comparative and International Education Society.

Grant Harman is an Emeritus Professor of Educational Management in the Centre for Higher Education Management and Policy at the University of New England in Armidale, New South Wales, Australia. His main research interests are in the academic profession and adjustment to the new commercial environment; research policy and research management at national and institutional levels; research commercialization and technology transfer; and international education. He has held academic appointments at the Australian National University and the University of Melbourne, and is currently the Editor in Chief of the academic journal, Higher Education, published by Springer in the Netherlands.

Jajah Koswara is a Professor in the Department of Agronomy, Bogor Agricultural University. She is former Director, Research and Community Service Development, Ministry of National Education; Vice-Chair, the University Research Council, Jakarta; Member of the National Research Council, Jakarta; Former Chair, the Governing Board SEAMEO, SEARCA; Member, the Oversight Committee of the McKnight Foundation Collaborative Crop Research Program, USA; Member of the Committee on Capacity Building in Science, International Council of Scientific Union, Paris; Former Member, the Task Force on Higher Education and Society, Harvard Institute for International Development, United States of America. She is a member, UNESCO Forum on Higher Education, Knowledge and Research.

Lynn Meek is Professor and Director of the Centre for Higher Education Management and Policy at the University of New England in Armidale, New South Wales, Australia. Trained in sociology of higher education at University of Cambridge, specific research interests include governance and management, research management, diversification of higher education institutions and systems, institutional amalgamations, organizational change, and comparative study of higher education systems. Professor Meek has published 27 books and monographs and numerous scholarly articles and book chapters. He is on the editorial board of several international journals and book series. He is a member of the UNESCO Scientific Committee for Asia and the Pacific.

Charas Suwanwela is Professor Emeritus and Chairperson of Chulalongkorn University Council, Bangkok; Chairperson of Policy Committee of Thailand Research Fund; Former President of Chulalongkorn University and of the Asian

Institute of Technology, Bangkok; Former Chairperson of National Higher Education Committee, Thailand; Former member of the UNESCO Advisory Committee on Higher Education and the UNESCO Preparatory Committee for World Conference on Higher Education; Chairperson of the Asia and the Pacific Follow-Up Committee for the World Conference on Higher Education. He is a member of the UNESCO Scientific Committee for Asia and the Pacific.

Muhammad Kamil Tadjudin is Professor of Biology in the School of Medicine, University of Indonesia, Jakarta, Indonesia. He was born in Jakarta and graduated from the Faculty of Medicine Universitas Indonesia in 1962. He was Rector of this University from 1994 to 1998. Since January 1999, he has been Chairman of the National Accreditation Board for Higher Education of Indonesia. He is a member of the UNESCO Scientific Committee for Asia and the Pacific.

Konai Helu Thaman is Professor of Pacific Education and Culture, UNESCO Chair in Teacher Education and Culture at the University of the South Pacific. Born and raised in Tonga, Konai did her undergraduate degree in New Zealand, and Masters and Doctorate Degree at the University of California at Santa Barbara and the University of the South Pacific. She has researched and published widely in the areas of curriculum development, culturally inclusive teacher education, women in higher education management, and indigenous education and research. She is also one of the Pacific's most prolific poets, with 5 collections of her poetry published, and several anthologized in books, such as Lali (1980) and Nuanua (1995), both edited by Albert Wendt. She is a member of the UNESCO Scientific Committee for Asia and the Pacific.

Wei Yu is Director and Professor of the South-East University in Nanjing, China. She has worked as a Professor, University President and Vice-Minister of Ministry of Education during the past 40 years. She is now the Director of the Research Center for Learning Science, Southeast University, China. Professor Yu is a prominent scientist and member of the Chinese Academy of Engineering. She is a former member of the UNESCO Scientific Committee for Asia and the Pacific.

Chapter One
Introduction

V. Lynn Meek and Charas Suwanwela

Introduction

A number of trends and issues have conspired to bring questions of "knowledge production," "research policy" and "research management" to the fore on the higher education policy agenda's of many countries. However, to date, analysis of the role of higher education in research and knowledge production has concentrated on North America and Western Europe. This volume is one of the first edited compilations to focus on different national perspectives on knowledge production and research in higher education in the Asia Pacific Region. There is very strong interest in this topic, both within the region and elsewhere, as the nations of the Asia Pacific Region build knowledge-based economies and in so doing, expand and adapt their respective higher education systems.

The book brings together leading experts on research and policy in higher education from the Asia Pacific Region. It arises from the meetings and deliberations of the UNESCO Forum on Higher Education, Knowledge and Research and the Regional Scientific Committee for Asia and the Pacific. The Forum focuses on higher education research and knowledge and provides a platform for researchers, policy makers, and experts to engage critically with research issues and research findings. Stressing the importance of strengthening research, particularly in developing countries, the Forum was established in 2001 within the World Conference on Higher Education follow-up framework in cooperation with, and funded by, the Swedish International Development Co-operation Agency (Sida). Its main objective is to widen understanding of systems, structures, policies, trends, and developments in higher education, research, and knowledge through the following: (i) gathering and engaging with existing and ongoing research; (ii) identifying research gaps and new priorities; (iii) stimulating and facilitating research; (iv) bringing current issues and debates to the fore; (v) making available research findings; and (vi) disseminating information on policies and practice (UNESCO, 2005). The Forum's activities are, in part, guided by the following four broad background issues affecting higher education and research nearly everywhere in the world today:

Globalization

Once viewed as mainly economic in nature, globalization has profound social and cultural aspects. Borders between countries have become more open to intellectual exchange, and the search for uniformity and common solutions continues to increase in many domains. In the field of higher education, numerous university activities involve international aspects as well. Many universities are part of international agreements, and mobility is facilitated by the rapid increase in international exchanges. In the field of research, there is an increased interest in the concerns of global governance, for example, democracy and human rights, collective social responsibility, the rising impact and interconnectedness of phenomena such as conflict resolution, multiculturalism, environmental matters, and the advent of technology. Significantly, there has been a noticeable emphasis on the need for research in the social sciences to fully understand the primary forces shaping the world today. (UNESCO 2005:1)

Transformation of Higher Education and Research

Today, student ranks are estimated at the 79 million mark and are expected to reach 100 million by 2025. Most of these will be in developing countries. This major change is forcing systems and institutions to diversify to meet the increasing demand for higher education. Similarly, the demand for research and knowledge is steadily increasing in the "knowledge-based society and economy." As a parallel trend, funding of higher education and research has not kept pace with the demand, and in many countries public investment in higher education and research has been declining. Higher education is also forced to compete with other government priorities besides that include other education areas. The result of this is that higher education institutions to an increasing degree seek funding from sources other than governments. Simultaneously, the private sector is taking a greater interest in the higher education system and its financing. This may be due to dissatisfaction with the type of training and skills the current system produces for the world of work or may stem from an interest in higher education based purely on a market logic. At the same time, the traditional hallmark of university research has been free enquiry and the ability to sustain long-term investigation. Yet, the present context tends to privilege other criteria, especially immediate application and diversified funding sources. As a result, a trend of "academic capitalization" or privatization of higher education and research has begun, whereby knowledge is viewed and treated as a commodity. (UNESCO 2005:1)

Research and Development (R&D) Trends

Worldwide, R&D trends vary considerably. While R&D has gained prominence in Japan in university and basic research, this same area has fallen sharply from its former level in Russia. Stability has remained in Scandinavia and the United States of America, but public support has dropped in countries where the funding role of government has been reduced. Finally, and perhaps the most worrying aspect, is the increased "targeting" of government support. This, when applied stringently, leaves little room for the flexible, long-term approach required for scientific research. Governments finance most of these activities in the Organisation for Economic Co-operation and

Development (OECD) countries, but in the developing world growing pressures to prioritize objectives mean that public funding for R&D work must be shared with other areas. A trend of stagnation or reduction of public funding of higher education and university research can be noted, as can a trend of increased private funding of research. It is important that governments continue to be the main source of funding both for university R&D and for university-based basic research as a part of this in order to safeguard the public good of research. Moreover, R&D is carried out in a variety of contexts (e.g., universities, public laboratories, and research centers, private nonprofit institutes and industry), and these tend to be fewer and weaker in developing countries. (UNESCO 2005:1)

Changing Patterns of Knowledge Creation

This trend has significant ramifications for university research and research-based teaching, as knowledge production and dissemination are increasingly being carried out in diverse contexts and through new media. So far, universities have excelled at generating knowledge. However, they have not yet proved their ability to reconfigure knowledge—that is, to draw creatively upon the entire distributed knowledge system, which is now much broader than the university sector alone. Old and new patterns can be noted. Older patterns can be described in terms of problems being context-specific and disciplinary, requiring homogenous skills, where hierarchical organization is respected and where knowledge stands alone and is evaluated by peer review. Newer patterns are described in terms of knowledge being produced in a context of application, transdisciplinary in nature, needing heterogeneous skills, organized around simpler and more temporary management structures, more socially accountable and reflexive, and more reliably assessed by a variety of practitioners (Gibbons, 1998). The shift in knowledge production is also affected by the advances in new information and communication technology (ICT), and it has obvious implications for research due to the expanding gap in ICT capacity between industrialized and developing countries. (UNESCO 2005:1).

The UNESCO Forum on Higher Education, Research and Knowledge consists of the following: (i) 5 Regional Scientific Committees; (ii) Global Scientific Committee; (iii) Co-ordination Committee; and (iv) Forum Secretariat. The Regional Scientific Committees comprise research experts and policy-makers for each of the five regions involved in the Forum initiative: Africa, the Arab States, Asia and the Pacific, Europe and North America, and Latin America and the Caribbean. Members of the Asia and the Pacific Committee are leading higher education experts from eight countries, many of which are the largest and most influential in the region: Australia, China, India, Indonesia, Japan, Oceania, the Philippines, and Thailand.[1] In this book, each Committee member analyzes the current knowledge production and research policy issues most pressing in the higher education systems of their respective countries. The book also draws on a number of commissioned papers and the engagement of Committee members with other higher education experts at a meeting held at the Indian Council of Social Science Research (ICSSR) in New Delhi, India, in September 2003 and the Committee's 1st Regional Research Seminar for Asia and the Pacific held at the United Nations University (UNU) in Tokyo in May 2004. The issues raised and lessons learnt from these meetings and papers serve as information for the country-specific case studies.

The book makes comparisons and contrasts and critically analyzes how policy in different countries of the Asia Pacific Region is encouraging a supportive (or otherwise) environment for the promotion of research within higher education. The book focuses on the capacity of higher education to promote social, cultural, and economic change through research. It examines recent changes that have taken place in the different Asian and Pacific countries' policies and practices regarding research in higher education in all of the broad disciplinary areas: sciences, social sciences, humanities, and the arts. The emphasis is on reviewing the changing context of governments' research policies and includes issues such as the impact of privatization, market forces, basic vis-à-vis applied/sponsored research, the relationship between teaching and research within and between higher education institutions, faculty development, equity and gender equality, and commercialization of research results. Research policy is broadly interpreted with the emphasis on understanding the different national policy contexts shaping how universities and other higher education institutions promote research for change, equity, and the social and economic well-being of society. The type of policies analyzed range from broad national policies on the funding of university research to institutional initiatives with respect to the management of intellectual property and knowledge transfer.

The remainder of this introduction will first summarize the key themes and issues resulting from the Committee's wide-ranging deliberations and consultations and will also explore the various perspectives and frameworks presented to it used to analyze the capacities of higher education institutions and systems to generate knowledge. The introduction concludes with a brief outline of what is to follow in the remaining chapters. The following discussion is adapted from the Rapporteurs' reports of the Committee meetings and of the Scientific Committee's 1st Regional Research Seminar for Asia and the Pacific, which took place in Tokyo, Japan, in May 2004, and the latter was recorded and transcribed.

Background Themes and Issues

As indicated above, this book is the product of the collective effort of the UNESCO Forum on Higher Education Research and Knowledge of the Regional Scientific Committee for Asia and the Pacific. Through an intensive process of meetings and seminars involving the presentation and exchange of papers amongst members and those of invited experts, the Committee has had the opportunity to consider and address many issues associated with higher education, research, and knowledge within the region. Some of the key issues are summarized below and are, of course, addressed in far more detail in the subsequent chapters.

The Committee held its inaugural meeting in Bangkok in February 2003 and it's second and more substantial meeting in New Delhi in September of the same year. The New Delhi meeting laid the foundation for the Scientific Committee's 1st Regional Research Seminar for Asia and the Pacific, subsequently held in Tokyo, Japan, in May 2004. The Regional Research Seminar was the genesis of this book. Participants at the New Delhi meeting and Tokyo seminar are listed in appendices 1 and 2.

The 2nd Meeting of the UNESCO Regional Scientific Committee for Asia and the Pacific, New Delhi, India, September 2003

The 2nd Meeting of the UNESCO Regional Scientific Committee for Asia and Pacific was held at the Indian Council of Social Science Research (ICSSR) in New Delhi in September 2003. At the New Delhi meeting, Professor V. R. Panchamukhi, chair of the ICSSR, in his welcoming address stressed the basic human liberating values of higher education, without which most, if not all, other functions of the sector are of little worth. The address also highlighted the importance of research funding coordinating agencies, like the ICSSR. As intermediary bodies among government, industry, and higher education institutions, research funding agencies play an important role in the scientific innovative framework of most nations. They do much more than simply distribute research funds. The ICSSR, for example, regards itself as a trendsetter in social science research and it helps focus research in three main areas: (i) relevance; (ii) usefulness; and (iii) sustainability. The importance of the policy role played by research funding agencies is an area deserving considerably more research. The point was also made that it is necessary to draw a distinction between knowledge and information. In many places it appears that what is developing is an "information society" rather than a "knowledge society." It is one of the key functions of higher education institutions to convert information into knowledge. Besides this values need to be brought back into the education process, without which education lacks substance and direction.

The Committee also received the presentation of a paper it commissioned from Professor Arun Nigavekar, chair of the Indian University Grants Commission (UGC), entitled "Trade in Higher Education" Nigavekar (2003). Higher education has itself become a tradable product and knowledge has become commodified. But as yet, there is little, if any, empirical evidence that GATS per se is compromising national systems of higher education. Trade in education should not merely be seen as the domination of developed countries over developing ones. In many instances, a more interactive relationship is evolving, particularly with respect to the GATS category of Commercial Presence. A country can simultaneously be a "sending" and "receiving" nation of educational services. Countries of the region that traditionally sent students overseas—particularly to Australia, the United States of America, and the United Kingdom—are now receiving foreign higher education students from the region and elsewhere. Proportionally, the largest recent growth in trade in higher education has been with respect to Commercial Presence, with universities in Australia, the United Kingdom, and elsewhere, establishing offshore programs, and sometimes full-fledged campuses, in various nations in the region. This may be done in partnership with a local university or private agent. The regulation of foreign universities operating domestically varies from country to country.

Much current analysis stresses that higher education institutions nearly everywhere are going through profound change and have, to a degree, lost control over their own destinies. Nearly everywhere, increasing demands are being placed on higher education institutions to address pressing national problems. In the context of the privatization of public higher education that is occurring in one form or another in most countries, research is not being prioritized or adequately funded. Many national knowledge systems are in need of change and reform. There is a need for greater flexibility in the way in which higher education systems are coordinated at the systems level, and critical decision making both within institutions and at the sector level needs to be better aligned with national priorities.

That said it also needs to be recognized that higher education institutions may be more resilient than commonly presumed and that they do much more than merely respond to external pressures. Higher education helps shape the very environment necessitating change, and in terms of theory, the relationship between higher education institutions and their external environment should be viewed as an interactive one. In some countries—Australia, Canada, China, Ireland, and so on—there has been significant real increase in research funding in the past decade and more firm recognition of the contribution of research to the "knowledge economy." While there are a number of issues in common across the various higher education systems, they must be interpreted in the context of local historical and cultural circumstances.

Developments in the area of higher education should not be viewed as either linear or inevitably moving in a certain predetermined direction. Many of the issues being discussed today have their historical antecedents, and over time, many of the same issues go on and come off the higher education agenda. Nonetheless, the complexity of the external environment with which higher education must engage has never been greater than it is at present.

Higher education needs to be viewed in terms of adequately balancing opposing forces, for example, technological change and preservation of cultural values, globalization and local interests, and training for generic skills and specialized knowledge. Possibly, there is need for developing a pragmatic education philosophy incorporating four key concepts: (i) Scholarship of Discovery, (ii) Scholarship of Integration, (iii) Scholarship of Application, and (iv) Scholarship of Teaching and Learning. However, insofar as Scholarship of Integration and Scholarship of Teaching and Learning assume a tight integration of teaching and research, it should be noted that there is considerable debate as to whether teaching and research are inseparable functions for all academics and all higher education institutions.

The content of university mission statements and their implementation is another area requiring more discussion and empirical investigation. Questions concern how university mission statements have changed over time and how much they differ from institution to institution and between national contexts within the region. It would be interesting to know to what degree mission statements are left at the rhetorical level or whether substantial effort is devoted to their actualization. At times there may be conflict between the core missions of particular higher education institutions and the core values of their academic staff. A related issue concerns the need to better understand the impact of globalization on higher education goals, philosophy, and mission. *Do mission statements take account of globalization? Are students being prepared for globalization and internationalization?* The ethics of developing certain types of knowledge, such as in the area of biological warfare, should be taken into account as well.

Clearly there is a need for more content-specific analysis of different national systems of higher education in the Asia Pacific Region. This involves the creation of new knowledge on the research process, taking into account indigenous knowledge systems amongst other things and the need to develop long-term strategies on the role of higher education in research and it's contribution to the "knowledge-based economy and society." However, much of the debate about the contribution of higher education to the global "knowledge-based economy" is Eurocentric and takes little account of indigenous knowledge systems and epistemology.

Alternative views and choices with respect to globalization require investigation. While nations should not regard themselves as powerless in the face of the forces of globalization, different and alternative ways in which national systems of higher education might respond to the forces of globalization are not yet well formulated or articulated. Some of the primary questions these issues raise include the following:

1. In a globalized world, what is the "appropriate" role of the state in providing education at the local level?
2. With respect to trade in education, what is the impact of new providers on the quality of education?
3. How does the increased marketization of higher education impact its governance and management?
4. What is the role of the university in the "knowledge-based society"?

It is these and related questions that have guided the Committee toward the preparation of this book.

Committee members have had the unique opportunity to exchange experiences based on the analysis of the contribution of their respective higher education systems to knowledge production. Some of the key issues and general lessons arising from this sharing of information and experience are summarized below.

Some countries in the Asia Pacific Region are in need of special advocacy and assistance. The Asia Pacific Region is vast with nations at various stages of development. Policies that might be applicable to a well-developed country like Australia might have little or no relevance to less developed countries. Some of the poorest countries of the region may not possess the resources or infrastructure to effectively play a role in a global "knowledge-based" economy, and special measures may be required to assist these countries.

The relationship between higher education and the labor market is a crucial one. Rapid expansion of higher education, as has occurred in China, in the absence of corresponding expansion of labor market opportunities creates substantial problems and distortions. Rather than tying expansion of higher education to increases in gross domestic product (GDP), an argument can be made that expansion should be tied to the size of the labor market and the need for highly trained graduates.

Mass higher education requires sector diversity. Not all higher education institutions can be elite research-intensive universities—no country can afford to fund all of its higher education institutions as elite research-intensive institutions. However, while there is a need for institutional diversity, there is also the corresponding problem of how to maintain both legitimacy and quality with respect to other types of higher education institutions that have more of a mass education focus.

One possible approach to strengthening national systems of higher education is through greater institutional and sector diversity. This is a very important issue, both educationally and politically. While the educational need for diversity may be demonstrated, there are often strong political forces requiring that all institutions are treated in the same way. Equality of opportunity in terms of participation and hierarchically segmented higher education systems may be in conflict, though the conflict has been addressed in several different ways in various countries.

Another important concern is the separation of teaching from research. As mentioned above, this is a hotly debated topic. Many people argue strongly that teaching in higher education institutions must be informed by research. However, the evidence suggests that nearly everywhere, even in research-intensive universities, a minority of academic staff produce the majority of research outputs (particularly with respect to externally funded research). The whole question of diversity and what type of role should be allocated to different categories of higher education institutions deserves much more serious consideration. In some national systems of higher education, however, it may be the case that the concept of the research university has little applicability.

The relationship between higher education, government, and the market is becoming more important and will continue to evolve, as will debates about the appropriate mix of public and private investment in higher education. It is recognized that there is a need to change thinking about higher education from regarding it in terms of immediate economic return to viewing it in terms of long-term sustainable development.

Market competition does not necessarily result in enhanced education quality. Government higher education policy-makers have been prone to assume that enhanced consumer choice and increased competition for students and scarce resources amongst higher education institutions will automatically improve quality. The evidence does not support such an assumption. Market competition can lead to an institution attempting to teach more and more students at a continuously declining unit cost, seriously compromising quality over time. It can also lead to institutions imitating each others products, thus decreasing the diversity of the sector as a whole.

On the other hand, market steering of higher education can be beneficial, particularly with respect to encouraging higher education institutions to diversify their funding base. The "marketization" of higher education has both positive and negative aspects. It is more on how market-oriented policies are applied in specific contexts than the "market" being all good or bad. More attention needs to be paid to the detail of the way in which market-oriented policies are formulated and implemented.

Generally, quality assurance mechanisms with respect to research are not well developed in many of the higher education institutions in the region. Traditionally, the quality of research was left mainly to the scientific community itself, with its emphasis on peer review. But the traditional approach to peer review has come under challenge, with governments intervening more directly in the research process through various policy initiatives and other mechanisms—such as the Research Assessment Exercise (RAE) in the United Kingdom. As mentioned previously, research funding agencies are playing a much more proactive role in assessing research quality and output. Also, more "outsiders," particularly representatives from industry, are being included in setting research priorities and assessing the relevance and quality of outputs.

In some countries in the Asia Pacific Region, there has been rapid expansion of private higher education, particularly private for-profit higher education. But some of these new institutions may be of dubious quality. Related to this is the need to think of strategies of how to better mobilize private investment in public higher education. There is a need to maintain an appropriate balance between the quality and quantity of higher education and, in so doing, to be cognizant of graduate employment opportunities.

Increasingly throughout the Asia Pacific Region, research management and control is becoming centralized, both within institutions and at the sector level. Governments are prioritizing their research funding, and institutions are concentrating their resources on a limited number of select areas of research, while concentrating control of research at the executive level. The impact of this on scientific innovation is not well understood and has the potential to stifle scientific creativity. The way in which research is organized and managed within higher education is a primary factor governing the relationship between research and the "knowledge-based economy." However, the impact of different management approaches on research productivity deserves much more serious investigation.

Many countries in the region are facing the question of the renewal of the academic profession as the present generation of academic staff reach retirement age. Related to this is the problem of attracting talented and well-trained young people into the academic profession. There is a growing amount of evidence to suggest that the present generation of graduates does not see the academic profession as an attractive career prospect, though there is a need for more empirical evidence on this issue.

Related to the career attractiveness of the academic profession is the problem of "brain drain," particularly when attractiveness is considered at the international level. While brain drain remains an important issue, it does not appear to be as great a problem as it once was, with the concept of brain drain being replaced by the notion of "brain circulation." Countries such as India, for example, have implemented policies to better utilize their nationals based overseas in local knowledge/scientific innovation networks.

Academic staff nearly everywhere are being asked to do more for less and be more productive. But this sometimes occurs in the context of inadequate faculty support. Higher education institutions may need to devote more time and effort to the development of their academic staff, particularly to development over and above traditional training and socialization in the disciplines. The modern academic engages in a much broader range of activities related to knowledge production than was the case 20 or 30 years ago. Many academics, for example, are being asked to be more entrepreneurial and to address complex issues associated with intellectual property rights (IPR). Teaching is another area where faculty development could be fruitfully applied. Still today in many instances, academic promotion is based more on research than teaching. Information and communications technology (ICT) is having a profound impact on higher education and is another factor that is making the role of academics more complex and difficult.

However, while the significance of the impact of new technology on higher education is undeniable, a few caveats are necessary. *First*, ICT should be viewed as a means to an end, rather than as an end in itself. *Second*, what may appear as leading-edge technology in one country may be conventional practice in another. *Third*, it does not appear that the virtual campus will replace the traditional one, and different types of higher education institutions using different technologies are not necessarily directly competing with one another—for example, in the United States of America the traditional four-year liberal arts colleges and virtual for-profit universities cater for distinct student clienteles. *Finally*, it was noted that application of ICT in higher education can increase the gap between the "haves" and the "have nots." For example, it may be inappropriate to put too much emphasis on ICT in situations where a substantial number of students do not even have access to electricity.

The above are some of the key issues identified by the Scientific Committee that are confronting the higher education institutions and systems in the region. In order to advance our understanding of these issues more systematic and detailed comparative studies, both historically and geographically based, would prove fruitful. Also, research on higher education policy could better inform policy makers than what it does and general higher education policy research could be better integrated with the more specialized area of science and research policy. Not enough is known about the differences and similarities of the higher education systems of the region, and there is a particular need to identify best practice in approach to particular issues amongst the various nations. In this respect, a key comparative research question could be formulated:

> What sort of enabling environment should governments provide in order to better stimulate higher education institutions' and systems' knowledge contribution?

Governments must have the political will to construct an appropriate environment in order for higher education-based research to flourish. Factors include the following: (i) roles and policies of funding agencies; (ii) articulation of national objectives; (iii) appropriate institutional incentives; (iv) innovation networks that incorporate business and industry as well as universities; (v) sustainability of funding; and (vi) wide-ranging public debate on the desired size, shape, and character of particular higher education systems. There is potential for greater interaction amongst the regional systems to address common problems and a need for cooperation as well as competition both *within* and *between* higher education systems and sectors. Approaches to a better understanding of these factors guided planning of the Scientific Committee's 1st Regional Research Seminar.

The 1st Seminar of the UNESCO Regional Research Scientific Committee for Asia and the Pacific, Tokyo, Japan, May 13–14, 2004 on "Changing Research Policy in the Higher Education Systems of the Asia Pacific Region"

The Scientific Committee's 1st Regional Research Seminar for Asia and the Pacific, which was held at the United Nations University in Tokyo on May 13–14, 2004, focused on changing research policy in the higher education systems of the Asia Pacific Region, emphasizing the capacity of higher education to promote social, cultural, and economic change through research. Chapters three to eleven of this book are revised versions of papers presented at the seminar and thus will not be summarized in this section of the introduction, though they are briefly outlined in the final section. A number of other key note addresses, discussants' responses, and significant interventions from the audience, however, were presented during the seminar and it is worthwhile to outline the main features of these below.

The Japanese Director General for International Affairs, Ministry of Education, Culture, Sports, Science and Technology (MEXT) Mr. Hiroshi Nagano welcomed the participants. Mr. Hiroshi Nagano maintained that the twenty-first is the era of knowledge. The growing demand for higher education, the rising use of ICTs in education and research, the increasing internationalization and demand for higher education in developing countries, and the growing number of new providers of higher education are just some of the developments that are changing the roles of traditional universities within higher education systems. With the transition to a "knowledge society," economic globalization, and the rapid development of science and technology, the importance of the role of higher education is increasing in all countries.

MEXT is in the process of creating an environment in which higher education institutions are able to operate more independently and autonomously. Institutions are encouraged to develop their own unique educational and research activities in accordance with their particular missions and objectives, and efforts are being made to build universities that are internationally competitive. Japan is introducing structural reform of universities through third party quality evaluations and by transforming national universities into independent corporations. Evaluation results will influence resource allocation, and third party evaluation has been introduced to guarantee the quality of universities and institutes of higher education. By becoming a university corporation, universities are expected to further develop their educational and research functions on the basis of their management, autonomy, and independence. The government has a responsibility to support national university corporations in terms of promoting academic research and producing professionals of the highest calibre. In addition, MEXT is promoting cooperation among the business, academic, and public sectors. It is important for Japan to promote business, academic, and public sector corporations to turn research achievements at universities and national research institutes into practical applications and thus return the benefits of this research to society.

Professor Hiroyuki Yoshikawa, president, National Institute of Advanced Industrial Science and Technology (AIST) spoke at the seminar about the "Japanese Policy of Science-Technology and Higher Education." Like many other countries, Japan is currently engaged in a process of research priority setting. The four broad areas of current interest are as follows: (i) life sciences; (ii) information/telecommunication technologies; (iii) environmental sciences; and (iv) nanotechnology and material technology. However, Professor Hiroyuki Yoshikawa cautioned against the naïve assumption that all society had to do was to invest in an area of research for substantial benefits to be realized. There is a time lag, sometimes a considerable one, between the creation of a new idea in science and its practical application. Professor Yoshikawa likened this to a sequence of "dream/nightmare/reality." The trick is to ensure adequate funding of innovative research during the "nightmare" period when the glow of initial great expectations has faded, but the reality of successful technological transfer has yet to occur. In Japan the field of biotechnology is in the period of a "dream"; researchers are enjoying substantial funding, as they keep ensuring society that in a very short time they will come up with tailor-made drugs and other great medicines. However this could actually take from 10 to 15 years, and the field of biotechnology will soon face the "nightmare" period. Others also made the point that the practice of research needs to be better understood by policy-makers. Research has its own saga, its own story, and this fact needs to be appreciated.

Professor Philip Altbach, director of the Centre for Higher Education, Boston College, spoke about "Winners and Losers in Asian Higher Education." The world in general, and Asia in particular, is at a turning point with regard to the role of research in higher education. Questions include: How to think about it; how to plan for it; how to do it? Many Asian countries are interested in, maybe even obsessed with, the idea of a "world-class" research university. But one must think carefully about *what is* world class—What is a world-class university and how can it be sustained? The future of the economies of some nations will depend on the decision taken concerning how research is to be organized, who is to do it, and what is its role in society.

Scientific progress in research is highly unequal—some countries have more, and some universities within countries have more. There are centers and peripheries in the scientific "knowledge system" on both the global and national levels:

> The fact is that the basic global status quo in international higher education is likely to remain for the coming several decades or more. The strength, size and focus of the major world higher education systems give them significant advantages—especially the USA. Asian countries have sufficiently deep problems [making it] difficult to see them emerging as the dominant academic systems—but having said that, it is likely that key Asian universities in the larger countries such as China, India, and Japan, and perhaps Indonesia, could emerge into the top rank but not as dominant institutions. Challenges include declining government support, privatization, pressures of enrolments in all Asian countries except Japan, academic traditions that do not fully support meritocracy and competition, in some cases overcentralization, language issues, and many others. (Altbach, 2004:1)

No nation can afford to fund all of its higher education institutions as if they were world-class research universities. Even in the United States of America, out of the 3,200 or so higher education institutions, only the top 100 receive 80 percent of the research funds allocated by either the federal government or private philanthropic foundations. Even in the United States of America the higher education system, as a whole, is not research-intensive.

Research-oriented universities have specific characteristics and requirements. These include the following:

- Full-time academic staff with doctoral degrees and a commitment to research. This might seem obvious, but many Asian universities lack . . . professors capable of doing research.
- Work responsibilities that recognize that research is part of the job—teaching loads that are not too high.
- The infrastructures at the university that will support research—libraries, Internet access, laboratories, supplies, equipment and the like. These facilities

must be kept up-to-date and similar to those found in the most advanced universities.
- Top quality students, especially at the graduate level.
- A research university must offer doctoral degrees and place considerable emphasis on graduate/professional degrees.
- Adequate financial support—including in all cases from governmental sources. Research universities can be private but nonetheless need governmental resources. Further, this support must be sustained over time. It is very damaging for support to vary considerably, as it does in many academic systems. The financial arrangements for a university can include tuition from students, support from private industry and others, external donors, and income from patents and consulting, but there must be a firm fiscal base as well.
- A clear vision of the goals of the research orientation—in most Asian countries and institutions, specific fields and departments will need to be targeted.
- Academic freedom and a culture of inquiry.
- The role of the English language (Altbach, 2004:3).

Only in the United States of America are there private research universities; no other country in any significant way has research universities in the private sector—it is simply too expensive.

Increasingly, there is an international market for academic and scientific talent in which the Asian countries must compete. And with all forms of market competition, there are "winners" and "losers":

> Inevitably, a limited number of Asian universities will be research-focused. Some countries will find it impossible to build up research capacity, and this must be clearly recognized.... Academic systems must be differentiated and there will be research "winners" and "losers" within countries as well. The academic profession itself will necessarily be differentiated. The development of research-oriented universities is not an easy task and there are many examples of failures or limited success in the USA and other countries. For Asia to compete in the "knowledge-based economies" of the 21st century, research universities and a research culture is necessary. (Altbach, 2004:2)

Professor Altbach made the point that most Asian countries have not made full use of universities and research to achieve their impressive levels of economic and social development. They have done it on the basis of things such as cheap manufacturing. But this is changing, which can be easily observed through the increased sophistication of the products, knowledge products in particular, coming from the Asian regions. Nonetheless, the academic systems need to catch up with the new economic realities, and those Asian countries that do not develop some kind of scientific infrastructure will be left behind in the economy of the twenty-first century. Asia is part of the world economy, and the world economy is knowledge-based. There are few countries that can hide behind cheap labor in the long run and prosper in this new economic environment.

In replying to Professor Altbach's presentation, Dr. Chan Basaruddin, Board of Higher Education, directorate general for Higher Education, Ministry of National Education (MNE), Indonesia, made the important observation that one of the principal drivers for universities to engage in research is the need and demand from industry and other institutions in the productive sector. The ability of universities to develop their research capacity will be determined, at least in part, by the maturity of the industries in their surrounding environment. In some developing countries there is little or no research undertaken or required by industry.

Academic systems in the twenty-first century will be differentiated academic systems, with a small number of research-intensive universities at the top. Difficult decisions need to be made in each country about how this differentiated system is to be organized, and funding for research will necessarily need to be concentrated at the top of the system. While this may sound elitist, it is the reality all nations must face, even wealthy ones.

Another point made during discussion was that Asia, as it thinks about its research culture and focus, needs to have the self-confidence that its ideas about research are as legitimate as anybody else's ideas about research. There is hegemony of knowledge and there is hegemony of ideas about the organization of knowledge and research, which is very much dominated by the West.

Professor Harman's paper is included in this book, and thus is summarized in the next section of this introduction. However, it is worthwhile to outline here some of the important points made by the discussant of Professor Harman's presentation, Dr. Allan Benedict Bernardo, academic vice-president of De La Salle University System, Manila, the Philippines.

Dr. Bernardo noted that developed countries have a number of policy instruments to apply to the management of research, such as priority setting and the competitive allocation of resources to research infrastructure: "These policy instruments are conceptualized and rationalized by the appropriate government agencies within well-defined although increasingly contested and evolving frameworks on the role of research in higher education and in attaining economic and social goals" (Bernardo, 2004:1). Government agencies have a number of ways and means available for fine-tuning policies on infrastructure, organization, and processes of higher education research. The availability of sophisticated policy instruments, however, is rarely a feature of developing countries that face challenges often substantially different from their more developed Asian neighbors. Higher education in most developing countries in the Asia Pacific Region was not founded on a research base. According to Bernardo,

> Typically higher education institutions were first established to provide post-secondary education for the socio-economic and/or intellectual elite. Eventually the higher education institutions and enrolments grew, although at different rates. . . . However, the growth of the higher education sector was focused on addressing the human resource development needs of these countries . . . [H]igher education was primarily designed to promote individual professional development and socio-economic mobility and it was possible to attain these goals without necessarily having a strong research base.
>
> In many developing countries in the region the infrastructure, organizational and work structures, incentive systems among others are not hospitable to research. For example, faculty member's work conditions typically involve heavy teaching responsibilities and research can only be undertaken over and above the teaching requirement. Libraries, physical facilities and other learning resources lean towards instructional

needs and do not support the specialized requirements for sustainable higher-level research programmes. Curricula emphasize the development and mastery of professional knowledge and skills instead of developing . . . analytic, critical, and creative thinking skills. Instruction emphasizes the consumption of research knowledge rather than problem posing and development of new knowledge. (Bernardo, 2004:1)

Policy-makers in many of the poorer countries of the Asia Pacific Region have had to concentrate on issues such as equity, access, quality, and efficiency rather than building a research base in higher education. Again, according to Bernardo,

[W]e can anticipate that higher education institutions in certain countries will be able to participate better in the more complex types of research activities compared to others. In these more developed systems the focus of research policy making is to fine-tune the existing priorities, structures and processes in order to better and more effectively participate in the global research enterprise and to address the emerging forces in the larger global environment. On the other hand, in the less developed systems the focus of research policy making is more developmental in nature. . . . The scenario might lead to a clear set of "winners" and "losers" in the arena of higher education research, where certain countries aggressively pursue their research development programmes while other countries for ever play "catch up." . . . We should pay attention to the concerns confronting the developing education systems, if we do not, there would emerge a clear divide between higher education systems within the region. (Bernardo, 2004:1)

The question is, however, How should the higher education systems in the developing countries of the Asia Pacific Region respond to these challenges?

A simplistic answer would be for these countries to radically and aggressively re-envision and re-engineer the higher education system in ways that give more emphasis to the research base of higher education. However, such radical movements are difficult to realize within higher education systems that are struggling with some fundamental problems of quality, access and efficiency. Externalities of these higher education systems would also most likely not be supportive of such radical changes.
A more realist response for developing countries would be to develop more strategic research policies that are rationalized within the diversity of higher education institutions of the country. As the higher education sectors in these countries are called on to address various concerns there might be a need for a policy environment within which different higher education institutions can develop to address specific concerns, leading to a diversification of universities. Probably in these countries most higher education institutions would not have the ability to develop the research culture or environment needed to effectively participate in the knowledge development process. However the state should create a higher education policy and regulatory environment, wherein selected institutions can grow to fulfil the more complex high-end research functions of a university. . . . [O]ne critical factor that would support such research policy initiatives in the developing countries would be strategic and collaborative programmes with more developed countries. Recent history has shown some outstanding examples of bilateral research collaborations between two countries of different levels of research capability. . . . As our countries strive to better develop policy instruments to enhance their respective higher education systems, we ought to discuss policy initiatives that would capitalize on strategic bilateral or multilateral collaborative efforts that have sustained long-term impact. (Bernardo, 2004:2)

In general discussion, it was also noted that in the Asia Pacific Region, the impression is that the leading nations are doing very well in terms of research policy and research and development. To a large extent they are following the same lines as many other OECD countries in that there is a greater role for the state, more public funding and effort to bring industry, universities, and public research institutes together in a new emphasis on research commercialization. In contrast, the higher education systems in the more developing countries of the region are in danger of becoming irrelevant to the increasingly globalized knowledge economy.

Outline of the Book

The issues raised above are further elaborated upon in the following two overview chapters (chapters two and three), which look at the impact of the rise of the global knowledge economy on countries of the Asia Pacific Region and assess key research policy issues associated with the changing role of the state and the effect on the region's higher education systems' capacity to promote research. William Cummings begins this review in chapter two. This author points out that knowledge utilization is nothing new in the Asia Pacific Region—for centuries, it has been central to the development strategies of the region. Countries like Japan have been most apt at incorporating Western knowledge and science while maintaining commitment to Eastern morality.

There are many scholars who maintain that the nations of the Asia Pacific Region, like developing countries everywhere, are in a peripheral position in relation to the Western nations, forced to accept knowledge and technological application produced elsewhere. Cummings, however, argues that the Asia Pacific Region is much more of a "knowledge production powerhouse" than commonly assumed and is steadily becoming more so. Many countries of the region are becoming key players in the global knowledge economy: China, India, Indonesia, Hong Kong, Korea, Malaysia, Singapore, Thailand, and Vietnam, to mention but a few. Cummings eloquently demonstrates that modern knowledge economies in the Asia Pacific Region have their roots as firmly in Eastern culture and tradition as in Western scientific practices. It is interesting to note that compared to Western countries, a larger proportion of financial support for research in the Asia Pacific Region comes from the corporate sector.

In chapter three, Grant Harman discusses national and institutional research policy for higher education in the Asia Pacific Region. Harman's chapter concentrates particularly on the topics of the role of the state in university research and development, the public funding of university research, priority setting, and university research links with industry and research commercialization. Research policy is defined as "guidelines and decisions expressed as directives, regulations or laws with regard to the funding, regulation, direction and monitoring of research activities." Harman notes that "research is defined differently, in different disciplines, in different countries, in different cultures. There are a number of different ways to classify research: basic, applied, curiosity or problem driven, etc."

Harman asks, "Why do universities engage in research?", and at the same time he replies that they undertake research to support teaching activity and particularly advanced-level research training and many academics have a strong commitment to conducting research. Research also is a very important service to society. Harman notes that higher education in the Asia Pacific Region reflects the tremendous diversity of this region. The higher education systems vary in size, resource capacity, student participation rates, research activities, and the mix of public and private. Many of these higher education systems are undergoing rapid change; just as the systems vary considerably so does the national research capacity. The strongest performers in terms of researchers per million populations are Australia, Japan, Korea, and Singapore and in terms of R&D expenditure Japan and Korea. The leaders in this region are amongst the leaders internationally.

Governments in the region are being forced to respond to pressures from diverse sets of stakeholders, especially from business firms and other consumers of research outputs. There are also changes taking place with respect to type of research activities, increased costs, and demands to capture research benefits. Governments are responding, argues Harman, in different ways. They are improving stakeholder involvement in priority settings with a much larger range of stakeholders being involved. They are restructuring research funding arrangements by redefining responsibilities of funding agencies, by combining agencies, and by developing new coordination mechanisms. Public funding has changed greatly over the past half century, from limited funding to support basic research and research training to an increased emphasis on mission-oriented funding and greater support for research and development. In many countries there is an increased emphasis on research commercialization. Harman concludes that fundamental economic and social changes are impacting significantly on higher education in the region and on national and institutional research policy.

The next eight chapters provide national perspectives on knowledge production and research in higher education in the following jurisdictions: Australia, China, India, Indonesia, Japan, Oceania, the Philippines, and Thailand. Lynn Meek (chapter four) opens the country reports with an analysis of policy and research management issues in Australian higher education. Though Australia has a small population base, it contributes about 3 percent to world scientific output. For the most part, research in Australia is primarily publicly sponsored. However, the nations' 37 public universities that enroll close to a million students are now largely privatized as government has reduced its financial support of higher education to less than 40 percent of the overall budget.

In terms of research policy at the sector and institutional levels, the emphasis in Australia has been on concentration and selectivity. Institutions have had to set research priorities in a national context where government also has set research priorities used to channel funding. Australian universities have developed elaborate research management structures, with all universities having a large research management office, headed by a professional research manager and overseen by an executive officer with a dedicated research portfolio.

Increasingly in Australia, research is being funded on the basis of outputs and the value of research defined in terms of its economic and commercial relevance. Though all institutions presently have a mandate to engage in research, most of the research funding and outputs are concentrated in a few research-intensive universities. The research teaching nexus is currently under question in Australia, with government introducing a research quality assessment framework, along the lines of the Research Assessment Exercise (RAE), United Kingdom, which will even more severely concentrate research funding in a few select universities.

Meek concludes that there is no one best approach to coordinating and funding university research at the national level. A number of competing demands must be balanced—balance and plurality are the key words. The public good nature of research must be recognized and supported. But as research becomes more elaborate and expensive, policies of concentration and selectivity are necessary. Governments and universities alike must make choices. But the choices must be informed ones—not driven primarily by ideology—and take place within a set of parameters that will sustain the research endeavor in the long term.

The recent and rapid expansion of higher education in China is on an unprecedented scale, as Professor Wei Yu demonstrates in chapter five. China has surpassed the United States of America in having the largest higher education system in the world. There are now nearly 20 million students and an enrollment rate of 17 percent of the of 18- to 22-year-old-age group. There are about 3,000 higher education institutions. Following the 1949 Revolution, China adopted the Russian higher education system, which locates research in separate academies rather than in universities. Following the 1979 reforms, universities began to engage in research as well as teaching. Research funding to universities has grown from CNY1.4 billion in 1990 to CNY21.9 billion in 2002. Currently, 60 percent of the scientific papers published within China and abroad are authored by university faculty. About 60 of the universities could be classified as research-intensive. Almost 80 percent of the postgraduate students are enrolled in these 60 universities.

Professor Wei notes that on the one hundredth anniversary of Peking University in May 1998, the then president Kiang Zein announced that China would build first-class world universities. The Ministry of Education embarked on the "985 Project," named after the date of President Kiang's announcement, establishing nine "first-echelon" universities, including Peking University and Tingha University. This was followed by support for more than 20 "second-echelon" universities to become leading world universities.

Half of the research funds going to Chinese universities is provided by the state and the other half by private enterprise. In China, being a developing country, the majority of the research is applied research with an emphasis on technological transfer. The establishment of science parks since 1999 has been one means for attempting to forge closer links between university-based research and the commercial sector. Some universities have established their own commercial branches, helping to facilitate technological transfer. It appears that the importance of the role of the university in contributing to the development of the "knowledge society" and "knowledge economy" is clearly recognized in China. An important characteristic of China is that many of the senior leaders have backgrounds in science and technology, particularly engineering. This may, in part, help explain the country's emphasis on research and technology in national development.

India has a large and complex system of higher education. Karuna Chanana notes in chapter six that until the early 1990s, Indian higher education was publicly dominated. Private institutions have been allowed to expand, but the system remains primarily public. From 1950 to the early 1990s, there was a phenomenal expansion of the Indian higher education system. In addition to universities, India has what are called "deemed to be universities"—institutions teaching single subjects, such as the Indian Institutes of Technology (IITs). There are also unitary universities that have no colleges affiliated to them, offering mainly postgraduate programs, and affiliating universities, such as the University of Delhi that has over 90 colleges under its umbrella. The Indian University Grants Commission (UGC) plays a critical role in setting standards and maintaining quality of the sector. In 2002, India had 288 universities: about 19 central universities, nearly 180 provincial universities, and about 76 mostly recently established deemed universities. There are a total of about 14,000 colleges. The affiliating colleges are mainly undergraduate colleges—90 percent of the undergraduate students are enrolled in them. The colleges are not regarded as research institutions. There are about 450,000 faculties in Indian higher education, of which 82 percent are in colleges and 18 percent in universities.

Soon after Independence in 1947, India framed a policy dictating that science will be promoted only in specialized institutions and independent research laboratories. India established the Council for Scientific and Industrial Research (CSIR), which allocated most of the funds to the specialized research institutions and little to the universities. Chanana points out that India does have world-class research institutions in the form of the All India Academy for Medical Sciences (AIIMS), the Indian Statistical Institute (ISI), Indian Institute of Technology (IITs), and others. These organizations are in the public sector and there appears to be little or no potential for private higher education institutions in India, or elsewhere in Asia, to emerge as an important part of the research enterprise.

The Asia and the Pacific Region is vast and diverse. It contains not only China and India with their huge populations but also the sparsely populated island continent of Australia with about 20 million people and the Indonesian archipelago with roughly 17,000 islands and a population of 250 million people and hundreds of ethnic groups and dialects. Enrollments in Indonesian higher education have grown from 1,000 students in 1945 to over 3 million students today. Jajah Koswara and Muhammad Tadjudin in chapter seven examine how the Indonesian higher education system is building a sustainable research culture. The issues to be addressed are substantial and include the following: (i) disparity in research capacity across universities and fields; (ii) poor research management; (iii) limited and unpredictable research funding; (iv) lack of an extensive research culture among academics; (v) low level of research quality; (vi) lack of a national research umbrella organization; (vii) low number of publications in national and international scientific journals; (viii) low appreciation of intellectual property rights; (ix) poor integration of research and graduate student research; and (x) limited collaboration with industry and international institutions.

Koswara and Tadjudin demonstrate that Indonesia is tackling these problems in various ways, particularly with respect to the nation's public higher education system. There have been significant policy shifts and innovations from the mid-1990s. The country has started a process of changing the system from a centrally controlled model, where universities were considered as part of the state bureaucracy, to a more decentralized system that enhances institutional autonomy. Long-term development strategies include institutional accreditation and evaluation, improving quality and relevance, enhancing access and equity, and the introduction of tiered competition that includes a number of quality enhancement schemes such as the University Research for Graduate Education (URGE). The goals of the ten programs that constitute the URGE program are to improve graduate programs, increase competitive funding for graduate education and university research, strengthen selection processes for grants and fellowships, integrate university research and graduate training, and attract highly qualified candidates for graduate education. With respect to their study of institutional capacity building, the authors draw a number of lessons. Tiered competition helps promote equity justice. The URGE project is excellent in terms of graduate capacity building. A research culture is beginning to emerge and is being integrated with educational programs, though publication in accredited scientific journals should be obligatory. Integration of research with community service is still difficult to achieve, and institutional incentives are still too low to prevent academic staff from "moonlighting." There are several types of research grants available to Indonesian academics with funding coming from a number of different ministries: grants for young researchers, research in gender studies, competitive research grants in natural science and technology, grants for fundamental research, grants promoting interuniversity research, grants for research in teacher education, grants for graduate students' research, and grants for research on classroom teaching in schools.

Japan like Australia is one of the highly industrialized countries of the region with a long tradition of investment in research and development. But as Akira Arimoto discusses in chapter eight, this country too is experiencing significant shifts in higher education policy, many of which impact on the conduct of research. Professor Arimoto first locates a discussion on Japanese research policy shifts in the broader context of general social change drivers, including globalization and the shift from what the author terms Knowledge-Based Society1 (KBS1) to Knowledge-Based Society2 (KBS2)—similar to (Gibbons et al., 1994) Mode 1 and Mode 2 science. Knowledge, social change, government higher education, and research policy are all interrelated.

Over the past decade, there have been many policy changes in Japanese higher education, including those that promote distinctive universities in a competitive environment, the promotion of science and research, and the incorporation of national universities, which began in 2004. Japan is attempting to bridge the gap between itself and the United States of America in terms of scientific and technological output. One option to achieve this end is the introduction of graduate schools along the lines of the North American model.

In chapter nine, Konai Helu Thaman argues for the inclusion of Pacific "indigenous knowledge systems" in the discourse on knowledge production and dissemination in higher education, particularly in higher education institutions in Oceania. Like indigenous peoples everywhere, the inhabitants of the islands of the Pacific Ocean have for centuries used local knowledge of themselves and their environment to live, work, trade, and communicate with one another. Western influence commencing about 300 years ago constitutes a small fraction of the thousands of years of history of these peoples. Thaman uses the term "indigenous knowledge systems" to refer to "specific systems of values, knowledge, understandings and practices, developed, and accumulated over millennia, by a group of people in a particular region, and maybe unique to that group or region."

"Indigenous knowledge systems" and "Western knowledge systems" are different but have equally valid ways of knowing and interpreting the world. Western knowledge claims universality, while "indigenous knowledge" is peculiar to the culture that owns it. In recent years there has been a concerted effort by some educators to incorporate "indigenous knowledge" into the formal education systems in Oceania, both to improve results and to preserve the cultural heritage of the Pacific people. Western scientific interest in "indigenous knowledge" is increasing. For some time, Western scholars have been interested in local agriculture and farming technologies. Presently, this interest has extended into the areas of environmental protection ("traditional ecological knowledge"), housing and health ("ethno-medicine"). But "indigenous knowledge" is more than making modern development more efficient and productive. It is, as Thaman argues, part and parcel of the "very identities and futures of Pacific Island people themselves." Nonetheless, "there remains a need to develop new methods (participatory, interdisciplinary research) to elicit and generate local knowledge, as well as innovative teaching methods that involve alternative forms of knowledge transfer, and to produce teaching materials that are adapted to local situations."

Many studies on education and educational reform have been conducted in the Philippines. Rose Salazar-Clemeña in chapter ten examines the development of national research policies and their institutional impact. The mandate of the Commission on Higher Education (CHED) is to formulate and recommend development plans, policies, priorities, programs, and research on and in higher education; recommend to government grants for higher education and research; develop criteria for allocating additional resources for research; and set research development priorities. CHED has produced a 10-year National Higher Education Research Agenda (NHERA) for the period 1998–2007.

The NHERA delineates policies, priorities, and procedures for the encouragement and support of research in higher education institutions in the Philippines. Its goals are to (i) push back the frontiers of knowledge in all the higher education disciplines; (ii) enhance instruction through strengthening bonds between teaching and research; and (iii) develop unifying theories or models that can be translated into mature technologies to improve the Filipino's quality of life. The policy attempts to ensure that the academic environment in higher education institutions nurtures and supports Filipino research talents and stimulates the development of a research culture. Given these policy directives, higher education institutions are expected to provide administrative support for research and to develop management capacities to help support research infrastructure. Several intervention strategies designed to increase the quantity and quality of research outputs of higher education institutions can be identified: (i) technical and financial aid to selected higher education institutions; (ii) linkages with foreign research institutions; (iii) support for research journals, awards, and other incentives; and (iv) development of research-oriented human resources through research training of promising junior faculty and graduate students. Research priority areas have been established based on the notion of multidisciplinarity.

Achievement of the NHERA goals is, not surprisingly, hampered by the meager funds that most institutions have available to allocate to research. The main problem with respect to boosting research productivity revolves around the research capacity of the higher education institutions. Of the 1,357 higher education institutions in the Philippines, 85 percent are private institutions relying mainly on tuition fees as their main source of income. Government funding to the state colleges and universities is spread ever more thinly as new public institutions are established. Thus, there is very limited financial support for research, and additional funding is urgently needed.

The last of the country case studies is Thailand presented by Charas Suwanwela in chapter eleven. Thailand is experiencing major higher education reforms, both structurally and functionally. The country is still in the massification phase of higher education expansion. There are about 1.7 million students enrolled in 126 Thai higher education institutions, 1 million of these in 24 public universities, including 660,000 students enrolled in the Open University. In 2002 there were around 220,000 students enrolled in 56 private universities. As is the case in a number of other developing countries, the expansion has mainly been in the area of the social science and humanities, where graduates experience high levels of unemployment. While the country has an excess of social science and humanities graduates, it has an insufficient supply of scientists, engineers, and health professionals. A national quality assurance framework for higher education has only recently been established.

The National Research Council (NRC) was established in Thailand in 1961 and the Thailand Research Promotion Fund (TRPF), the National Science and Technology Development Agency (NSTDA), and the Health Systems Research Institute (HSRI) were created in 1993. The National Science and Technology Development Agency (NSTDA) has three centers: National Electronic and Communication Technology Centre, National Biotechnology Centre, and National Metallurgy and Material Centre. The NSTDA also runs Thailand's science parks, which involve some academics. The Thailand Research Promotion Fund (TRPF) is solely a grant-giving agency. Only about 17 percent of the total budget for research goes to the universities, which are only a small component of the research and development field.

According to Suwanwela, the relatively low investment in higher education that has prevailed for several decades has been one of the root causes of the problems faced by the Thai higher education sector. Higher education institutions must depend more and more on sources of income other than the government. There is a need to diversify research funding in the universities. The Thai case clearly demonstrates the importance of the networking of researchers in the various agencies, including universities.

In chapter twelve, the concluding chapter, Meek attempts to bring together the various arguments presented in this volume to form a comprehensive final analysis of the role of higher education in "knowledge production" in the Asia Pacific Region. Meek questions some of the common assumptions underpinning recent arguments concerning the transformation of higher education and its role in the "knowledge economy and society," such as the inevitable march toward the introduction of New Public Management principles in higher education and the transition of science from a traditional disciplinary base to diffusion throughout society. This chapter suggests an agenda for future research on higher education and knowledge production in the Asia Pacific Region.

Appendix 1: The 2nd Meeting of the UNESCO Regional Research Scientific Committee for Asia and the Pacific, New Delhi, India, September 2003

The following participants attended the Regional Scientific Committee Meeting for Asia and the Pacific, which took place on September 8–9, 2003 in New Delhi, India.
Welcoming Address:

- V. R. Panchamukhi—Chair of the Indian Council of Social Science Research (ICSSR)

Committee Members:

- Akira Arimoto—Chair (Japan)
- Karuna Chanana (India)
- Konaiholeva Helu-Thaman (Fiji)
- V. Lynn Meek (Australia)
- Rose Marie Salazar-Clemeña—Vice-Chair (the Philippines)
- Charas Suwansela (Thailand)
- M. K. Tadjudin (Indonesia)
- Yu Wei (China)

Author of a Commissioned Paper:

- Arun Nigavekar—Chair, University Grants Commission (UGC), New Delhi, India

UNESCO Staff:

- Qutub Khan (New Delhi)
- Terry Tae-Kyung Kim (Paris)
- Katri Pohjolainen Yap (Paris)

Appendix 2: The 1st Seminar of the UNESCO Regional Scientific Committee for Asia and the Pacific, Tokyo, Japan, May 2004 on "Changing Research policy in the Higher Education Systems of the Asia Pacific Region"

The following participants attended the Regional Research Seminar for Asia and the Pacific, which took place on May 13–14 2004, in Tokyo, Japan.
Welcoming Address:

- Hiroshi Nagano—Director General for International Affairs, Ministry of Education, Culture, Sports, Science and Technology (MEXT)

Committee Members:

- Akira Arimoto—Chair (Japan)
- Karuna Chanana (India)

- V. Lynn Meek (Australia)
- Rose Marie Salazar-Clemeña Vice-Chair (the Philippines)
- Charas Suwansela (Thailand)
- M. K. Tadjudin (Indonesia)
- Yu Wei (China)

Keynote Speakers:

- Philip Altbach (United States of America)
- Grant Harman (Australia)
- Hiroyuki Yoshikawa (Japan)
- Jajah Koswara (Indonesia)

Discussants for Keynote Address:

- Chan Basaruddin (Indonesia)
- Allan Benedict Bernardo (the Philippines)
- Jandhyala B. G. Tilak (India)

Participants:

- Richard Braddock (Australia)
- Do Van Xe (Vietnam)
- Fumi Kitagawa (Japan)
- Futao Huang (Japan)
- Hans Van Ginkel (Japan)
- Hiroshi Nagano (Japan)
- Li Zhi Min (China)
- Evangelia Papoutsaki (Papua New Guinea)
- Shuji Uchikawa (Japan)
- Norietta C. Tansio (the Philippines)
- Jean C. Tayag (the Philippines)
- Ulrich Teichler (Germany)
- Utak Chung (Republic of Korea)

UNESCO Staff:

- Yonemura Akemi (New Delhi)
- Molly N. N. Lee (Bangkok)
- Tony Marjoram (Paris)
- Min-Chul Shim (Paris)
- Tawfik Mohsen (New Delhi)
- Katri Pohjolainen Yap (Paris)

Note

1. In 2005, membership of the Asia and the Pacific Committee was extended to Malaysia, Mongolia, and South Korea. However, these countries are not included in this book.

References and Works Consulted

Altbach, Philip (2004) "Winners and Losers in Asian Higher Education: The Strengthening of a Research Culture—Abstract." Paper presented at the first Seminar of the UNESCO Regional Research Scientific Committee for Asia and the Pacific, Tokyo, Japan, May 13–14, 2004.

Bernardo, A. (2004) "Discussant in Reply to Professor Grant Harman." Paper presented at the 1st Seminar of the UNESCO Regional Research Scientific Committee for Asia and the Pacific, Tokyo, Japan, May 13–14, 2004.

Gibbons, M. (1998) "Higher Education Relevance in the 21st Century." Human Development Network, World Bank, Washington, DC.

Gibbons, M., C. Limoges, H. Nowotny, S. Schwartzman, P. Scott, and M. Trow (1994) *The New Production of Knowledge: The Dynamics of Science and Research in Contemporary Societies*. London: Sage Publications.

Nigavekar, A. (2003) "Trade in Higher Education: The Impact of the General Agreement on Trade in Services (GATS) on Higher Education, Research and Knowledge Systems in Selected Contexts in Asia and the Pacific." Paper presented at the 2nd meeting of the UNESCO Regional Scientific Committee for Asia and Pacific, New Delhi, September 8–9.

UNESCO (2005) "UNESCO Forum on Higher Education, Research and Knowledge: Some General Trends and Challenges," UNESCO, Paris, http://portal.unesco.org/education/en/ev.php-URL_ID=21052&URL_DO=DO_TOPIC& URL_SECTION=201.html (accessed July 31, 2005).

CHAPTER TWO

MODERNIZATION, DEVELOPMENT STRATEGIES, AND KNOWLEDGE PRODUCTION IN THE ASIA PACIFIC REGION

William K. Cummings

Introduction

Knowledge utilization has been a core element in development strategies since the dawn of modernization in the Asia Pacific Region. An American expedition under Commodore M. C. Perry shocked the Tokugawa shogunate in 1853 by entering Japan's Edo Bay (now Tokyo Bay) with smoke-belching steam-driven warships and later by Perry's demonstration of various other technologies, including a miniature train. Japan's leaders realized that their closed society was vastly inferior in military capability to the Western Barbarians; and unless they caught up they were destined for colonial subjugation much like what was taking place across the Sea of Japan, in mainland China. Thus, within a few short years a new leadership emerged in Japan, which declared strong determination "to seek knowledge throughout the world" and to accept Western science—at the same time as they reaffirmed Eastern morality (Bartholomew, 1989).

Substantial investments were devoted from the earliest days of the Meiji Revolution (1869) for enhancing Japan's understanding and utilization of science and technology as a means of strengthening the nation. At first the Japanese focus was on "knowledge imitation." A new institute was established to translate foreign knowledge and other new institutes specialized in engineering, ship building, armaments, and other technological areas; subsequently several were consolidated in the Tokyo University—which was in 1886 renamed as the First Imperial University. Over the next 30 years, numerous other public and private higher education institutions were founded, most with a focus on Western science, technology, law, and languages. By the 1920s increasing emphasis was placed on "knowledge innovation," and from the 1970s Japan began to place a stronger emphasis on "knowledge creation," (Cummings, 1990). Some of the themes underlying this shift were drawn from the West and especially from the United States of America. But as will be argued below, Japan has also fostered some new strategic directions (Kodama, 1991).

Over time, and especially over the past 3 decades, other Asia Pacific societies have like Japan taken bold steps to accelerate the processes of "knowledge innovation and creation." Korea, Singapore, and Taiwan were most notable for their bold steps over the past decade or so, but the trend is evident throughout the region. Each nation faces its unique set of *opportunities and obstacles that we also acknowledge*. One obstacle frequently cited is the supposed Western, and especially the USA, dominance of "global knowledge production" so, according to this view, the West usually makes discoveries first and similarly is more efficient in translating its basic discoveries into applications; thus Asia is said to be locked in a peripheral or semicore position in global knowledge production (Altbach and Umakoshi, 2004; Marginson, 2004). While recognizing the obstacles, we will argue that the region has much more potential than is generally appreciated—investment, talent, unique biosphere, humanistic objectives, and a collaborative spirit—and an impressive array of recent accomplishments. This suggests the prospect that the Asia Pacific Region may be emerging as a *new powerhouse of knowledge production*.

The Context

Before considering recent trends in development strategies, it will be useful to highlight several relevant characteristics of the region:

A Rich and Distinctive Intellectual Tradition

The Asia Pacific Region is both the sight of some of the world's greatest civilizations that have in past times added immensely to the world's stock of knowledge, and of some of the world's most primitive peoples. India has given birth to the great religions and philosophies of "Hinduism" and "Buddhism" that include profound insights into the nature of the Cosmos, and China is the home of the "Confucian political and social philosophy" as well as an extraordinary tradition of scientific and technological discovery that superseded the accomplishments of the West at least throughout the sixteenth century (Needham and Ling, 1956).

The strong intellectual traditions of these two civilizations provide an important part of the base for contemporary developments. As Nakayama (1984) observes, "Asia in these early times developed a distinct mode of inquiry, the documentary tradition, which stands in sharp contrast to the Western rhetorical tradition." The documentary tradition trains the mind to build a strong foundation in basic principles, to carefully assemble all of the relevant information, and to take small first steps in discovery as the foundation for a later stage of boldness. The subsequent exposure to Western modes of inquiry complemented the Asian documentary tradition.

Colonialism Stunted the Development of Educational Development and Knowledge Production

Whereas major civilizations and large societies prevailed in China and India, in other parts of the Asia Pacific Region, notably Oceania and to a lesser degree in the areas now known as Indonesia and the Philippines human settlement was sparse, social organization simpler, and the practices of writing and recording very limited. For example, the major empires of Indonesia and mainland Southeast Asia largely borrowed their social and political theories from the cultures of China and India.

The cultural and scientific development of much of the Asia Pacific Region was punctuated by the arrival of Western colonizers and settlers who set about introducing a new layer of externally oriented institutions on "old" societies. The primary focus of the Western invaders was on the exploitation of agriculture—silk from China, tea from India, spices from Micronesia and Polynesia.

In order to advance these extractive goals, the colonizers and pioneers set up minimal education systems leading in most cases to a handful of higher education institutions focused primarily on law and the humanities, fields that were believed appropriate for the development of civil servants. In some locations, fledgling institutes for the study of agriculture and the biosphere were also established—for example, Raffles initiated the Botanical Gardens at Bogor, Indonesia—but in general knowledge production was not given much consideration.

Asia Pacific States Treasure their Autonomy

With the end of World War II, the colonial powers began to depart from the Asia Pacific Region and there ensued a period of political consolidation. The Maoist victory in China was the first step with the Kuomintang government exiting to Taiwan. From the early 1950s, nationalist guerrillas began to mount their struggle against France and later against the United States of America.

The process of state formation led to the emergence of societies that varied widely in terms of ethnic-cultural diversity. For example, India and Indonesia both include many religious and national-ethnic groups while on the contrary Japan and Korea are somewhat more homogeneous. In between are nations such as Malaysia and Thailand that favor one group by stressing the cultural assimilation of their minority groups. Occasionally the cultural differences within particular Asian nations become a source of conflict as in the 2004 protest of the Muslim minority in southern Thailand. When domestic tensions appear in an Asian nation, most Asian nations view this as an "internal matter" and restrict their criticism. Myanmar's neighbors have tolerated its repressive system for decades without exerting notable pressure for reform.

During the late 1950s and early 1960s, tensions flared between Indonesia and its neighbors, and Malaysia also experienced a communist incursion. Thus the region has experienced considerable tension and periodic conflict. As most of the Asia Pacific countries have, in relatively recent times, had to defend their boundaries against outside incursions, they are wary of foreign penetration.

This wariness about foreign "political" penetration extends to Asia Pacific views on foreign "economic" penetration. Most of the states of the region have a history of setting up barriers to unwanted penetration of their economies by foreign investment or imports. While South Korea accepted large loans from the World Bank (WB) in the early decades of its development, it later placed high priority on paying back these loans and observing clear limits on foreign indebtedness (Stallings, 1990). China until recently did not accept WB loans or foreign investment; while China's policy has seemingly radically changed over the past decade, it is nevertheless the case that Chinese firms usually maintain a *controlling interest* in partnerships that involve foreign investment. Looking across the Asia Pacific landscape, perhaps only Indonesia has allowed itself to be seriously overexposed by foreign investment.

Asia Pacific States Place a High Priority on Economic and Social Development

Partly as a result of the postcolonial history of political struggle, many of the Asia Pacific nations emerged with strong states that were accustomed to making the major decisions on the future directions for national development. Some observers refer to the Asian pattern of politico-economic organization as the "Development State" (Johnson, 1982), implying strong leaders, a single party, a high commitment to economic development, and a minimal commitment to democracy. While it cannot be said that the structure of the Asia Development State provides the explanation, it nevertheless is noteworthy that several of the Asian countries have been exceptionally successful in promoting economic development with equity. A World Bank (1992) study highlighted the success of Hong Kong, Korea, Singapore, and Taiwan referring to these as "miracle" economies. The study also suggested that China, Indonesia, Malaysia, and Thailand were near miracles. Since that time Vietnam has begun to show promise, as have parts of India.

Over time, several of the Asia Pacific states have become more politically inclusive, though usually within a framework of firm political leadership focused on economic development. Increasingly, these states have beamed in on knowledge production as an important key toward promoting national development. Of course, the differences in context outlined above have influenced the respective approaches to knowledge production.

Asia Pacific States View Human Resources as the Foundation of Development.

Most Asia Pacific states recognize the importance of a well-educated population for the realization of development goals, and thus stress universal basic education (UBE) of high quality with considerable opportunities for further education up through graduate studies. In most Asia Pacific school curricula, science and mathematics are featured from the earliest grades, and as demonstrated repeatedly in international studies of academic achievement young Asian people do exceptionally well; for example, in the "Third International Mathematics and Science Survey (TIMSS)," the average achievement scores of young people from China, Hong Kong, Japan, Korea, and Singapore were ranked at the very top among some 40 countries (IEA in NSB, 2004:1–13). Science and mathematics is featured in the secondary and tertiary levels of Asia Pacific education with the result that China, India, and Japan graduate a larger number of first-degree holders in science and engineering than does the United States of America or Russia, not to speak of the Western European countries. The strong foundation in human resources means that the Asia Pacific research and development (R&D) enterprises have a substantial reserve of candidates when they seek to staff new entities.

Asia Pacific States Vary in their Development Priorities

Virtually all of the Asia Pacific nations place a high priority on self-sufficiency and thus have, at least in the past, placed much emphasis on improving the quality and efficiency of their agricultural production. Several nations continue to emphasize agricultural exports as a major component of their national revenues. However, many Asia Pacific states have high population densities and labor costs that strain their potential for further gains in agricultural productivity, and thus they have elected to emphasize manufacturing and the services as current and future areas of economic growth. With the stress on manufacturing and service, each nation has choices concerning particular industries to emphasize whether the focus should be on world-class cutting-edge products or the more efficient production of familiar products. The respective choices have clear implications for national science and technology policies.

Defense-Related Knowledge Production Is Not a Priority

While the region has a history of conflict, especially over the past 2 decades the level of conflict has considerably subsided. Regional tranquillity has been realized, at least in part, because of regional dialogs fostered by organizations such as the Association of Southeast Asian Nations (ASEAN), Asia-Pacific Economic Cooperation (APEC), and the Economic Commission for Asia & the Far East (ESCAFE). Thanks to regional tranquillity, most Asia Pacific regions devote relatively modest amounts of their national budgets to defense budgets and to defense-related research and development. Whereas in the United States of America and Western Europe, upward of one-third of a nation's R&D expenditures might focus on defense, the typical proportion in the Asia Pacific Region is one-tenth, leaving much greater scope for commercial and academic R&D.

The Scale of Asian Nations Varies

Asia Pacific nations vary immensely in geographic scale from large countries such as Australia and China on the one hand to a small country such as Singapore on the other. Of even greater importance for the execution of research and development programs are the wide differences in demographic scale: without a critical density of researchers in a particular area of inquiry, it is difficult for a nation, on its own, to foster major discoveries in research and development. To a certain degree, a high allocation of resources can compensate for a small scale as is demonstrated by Finland and Switzerland and in the Asia Pacific Region possibly by Singapore. Also, small scale leads a nation to buy brains (expatriate researchers) and ideas (technology licensing) alongside energetic efforts at homegrown science and technology. Even so, large nations such as China and India have a natural advantage, as the sheer human scale of their research and development enterprise enhances the probability of identifying native talent and nurturing homegrown discoveries.

New Focus on Knowledge Creation

For most of the past century knowledge production was centered in the West, and other regions of the world including the Asia Pacific Region sought to draw on Western knowledge in order to "catch up." Into the 1970s, this strategy was clearly evident even in the case of Japan, the region's most technologically advanced society. For example, Japan's early successes in textiles, steel, automobiles, electrical and electronic goods were largely based on the application and refinement of imported technology.

However, from at least the late 1960s, Japanese policy-makers came to recognize that Japan was pressing on the upper edge of imported technology utilization and thus that the future prospect for low-cost borrowing technology was bleak. Thus, it would be necessary for Japan to place increasing emphasis on the autonomous development of technology. Just as Japan began to make this policy shift, over the next 2 decades other Asia Pacific nations came to the same conclusion: Korea and Taiwan in the mid-1980s; Australia, Singapore, and Malaysia in the early 1990s. An example is Malaysia's vision 20–20 (Sarji, 1993) that, among other innovative concepts, proposes the development of a new information highway and to that end a range of new programs aimed at fostering a homegrown creation of a wide range of information technologies.

The new focus on knowledge creation is accompanied by increased funding for research and development. While in the 1960s, Japan was devoting only about 1 percent of its gross national product (GNP) to research and development (R&D), this was doubled by the early 1980s and has continued to rise since then. In 2001, it was 2.98 percent or fourth in world. In that same year, the average expenditure for R&D of OECD countries was 2.24 percent, and in the United States of America it was 2.71 percent. Among other countries in the Asia Pacific Region, Korea's expenditure for R&D had risen to 2.65 percent, Singapore to 2.11 percent, Taiwan to 2.05 percent (only civilian R&D), and Australia to 1.53 percent. Several other countries in the region devote upward of 1 percent to research and development (NSB, 2004:4–51).

The Purpose of Science and Technology

From the earliest days of Japan's Meiji Era (1868–1912), increased knowledge of Western science was seen as a means toward *increasing national strength* in the face of possible Western domination. Japan, avoiding colonization, rapidly became a significant world power and increasingly an aggressive one controlling China in 1894 and gaining victory over Tsarist Russia in 1904–1905. While Japan assumed a minor role in the World War I, in the ensuing years it declared a "Greater East Asia Prosperity Sphere" and proceeded to conquer much of East and Southeast Asia, and as well many Pacific Island groups during the World War II. Science, including academic science, was mobilized for Japan's militaristic expansion, but this aggressive push was ultimately concluded by a science-based response: the horrific USA atomic bombs dropped on Hiroshima and Nagasaki (August 14, 1945) leading to Japan's unconditional surrender. With Japan's defeat, the Japanese people concluded and wrote into their new constitution that they wished to have no more involvement in war. And Japan's academic establishment expressed its shame: it's contribution to the wartime

Table 2.1 Distribution of government R&D budget appropriations in selected countries by socioeconomic objective: 2000 or 2001

Socioeconomic objective	United States-2001	Japan-2001	Germany-2001	France-2000	United Kingdom-2000	Russian Federation-2001	South Korea-2001
Total (millions of USA dollars)	86,756	23,153	17,946	14,605	10,030	5,889	6,195
1. Exploration and exploitation of the earth	1.2	1.9	1.8	0.8	1.3	1.5	1.5
2. Infrastructure and general planning of land use	2.0	4.4	1.7	0.6	1.2	1.2	4.2
3. Control and care of the environment	0.7	0.8	3.1	2.9	1.6	1.6	4.5
4. Protection and improvement of human health	24.8	3.9	4.0	5.8	14.6	2.0	7.1
5. Production, distribution, and rational use of energy	1.5	17.4	3.4	3.9	0.5	2.0	4.7
6. Agricultural production and technology	2.5	3.5	2.4	2.1	4.1	9.9	8.4
7. Industrial production and technology	0.5	7.5	12.1	6.3	1.7	11.4	29.5
8. Social structures and relationships	0.9	0.9	4.5	0.8	4.1	2.0	2.6
9. Exploration and exploitation of space	7.1	6.7	4.7	9.8	2.2	10.1	3.2
10. Research financed from GUF*	N.A.	34.8	39.0	21.6	19.6	N.A.	N.A.
11. Nonoriented research	18.5	13.8	16.1	19.8	12.1	14.0	15.8
12. Other civil research	6.3	0.0	0.1	2.3	0.3	0.9	0.0
13. Defense	52.7	4.3	7.1	23.2	36.6	43.5	15.8

Notes: N.A. = not available.

* United States of America, Russian Federation, and Korea do not have a category equivalent to General University Funds (GUF).

Conversions of foreign currencies to USA dollars are calculated with Organisation for Economic Co-operation and Development (OECD) purchasing power parity exchange rates. Percentages may not sum to 100 because of rounding. USA data are based on budget authority. Because of General University Funds (GUF) and slight differences in accounting practices, the distribution of government budgets among socioeconomic objectives may not completely reflect actual distribution of government-funded research in particular objectives. Japanese data are based on science and technology budget data, which include items other than R&D. Such items are a small proportion of the budget; therefore, data may still be used as an approximate indicator of relative government emphasis on R&D by objective.

Source: National Science Board. Science & Engineering Indicators—2004 from OECD, unpublished tabulations (Paris, 2003); and OECD, Main Science and Technology Indicators (Paris, 2002).

effort. Hence, for the future Japan declared that science should be for peace and not war, for the people and not the leaders.

Out of this sober reflection, Japan began to envision a new role for science involving not only the economic prosperity of the nation but also the improvement of the natural and social environment. This vision has been reflected in the subsequent development of Japanese science and technology policy. Official descriptions of Japanese science and technology policy are notable for their humanistic emphasis on topics such as environmental preservation, improving the quality of urban life, and creating a more comfortable setting for older people. As noted in table 2.1, public funding of research is substantial in all countries tending to average about one-third of all funding, but the government's proportion of funding is largest in the United States of America primarily due to the USA government's substantial commitments for defense-related research. Government's share is somewhat less in the Asia Pacific Region.

The allocations of government S&T resources by purpose in Japan, as reflected in table 2.1, place far less emphasis on defense-oriented science than does the United States of America or the United Kingdom and far more on other areas such as energy, industrial applications, planning of land use, and university research (the funds in the general university funds and nonoriented research categories). The allocations in South Korea, the only other Asia Pacific nation for which comparable data is available, tend to follow the same pattern as Japan—relatively small allocations on defense, more on civilian priorities including agriculture and land use, and university research.

A Distinctive Strategy or Strategies for Knowledge Creation?

While science and technology have played a major role in the development of nations for several centuries, it is only after World War II that the major industrial nations, led by the United States of America, began to develop coherent science and technology policies. Vannevar Bush (Department of Electrical Engineering, MIT) then vice-president and dean of MIT, and scientific advisor to the then president of the USA, observed that "there is a perverse law governing research: Under the pressure for immediate results, and unless deliberate policies are set up to guard against this, *applied research inevitably drives out pure*. The moral is clear: It is pure research which deserves and requires special protection and specially-assured support" (Bush, 1945:83).

Bush and his colleagues depicted a *linear model of knowledge production* with basic research as the foundation generating fundamental breakthroughs that would foster applications that could then be developed into new products and services. One outcome in the United States of America was the establishment of the National Science Foundation (NSF) in 1950 and the National Institutes of Health (NIH) in 1948 as federal government sources for basic research funds that distribute these funds to capable scientists on the basis of peer-reviewed evaluations of their research proposals. In the years that were to follow, basic science was strengthened in the United States of America, especially in the top strata of higher education institutions that came to be known as research universities. Additionally, the USA federal government came to play a prominent role in the support of applied and development research in laboratories of private industrial firms. Thus, the science and technology model pioneered by the United States of America stressed strong support for basic research and a substantial role for the federal government in the support of both basic and applied research.

While the USA model was able to "leapfrog" American science into a leadership position in basic science in the postwar period, few other governments had an equivalent level of resources for the actual funding of research. Rather in other settings the government decided to limit its role to serving primarily as a facilitator of research through providing information and offering tax and tariff incentives while looking to other sources, notably the private sector for funding. This pattern was particularly noticeable in Japan and since then in many of the other Asia Pacific nations. For example, while in the United States of America in 1985 nearly 40 percent of all research and development was supported by the federal government, the Japanese government only funded 22 percent of all Japanese R&D. Over the past 2 decades, there has been a modest convergence with the USA government's share of funding decreasing to 35 percent and the Japanese government's share increasing to 25 percent. But the basic contrast persists. The Japanese pattern of a greater reliance on commercially funded research is also found in Korea, Singapore, and Taiwan.

The Asia Pacific emphasis on applied research, and a larger role for the commercial sector in research and development, implies a distinctive approach sometimes referred to as the *interactive model of knowledge production*. In the interactive model, each sector has a substantial role in research and development, and each sector devotes at least some effort to all phases of the R&D continuum from basic to developmental research. Also, while the "linear model" assumes that basic research is the source of new research directions, in the "interactive model" it is acknowledged that important new research directions may be suggested as researchers discover shortcomings in their applied and developmental research. Rather than a unilinear conception of the R&D endeavor the interactive model makes no assumptions about directionality.

The Role of the Universities

Depending on the model, the role of the universities differ. In the linear model, the university has a prominent role in basic research and human resource development. Because of the university's considerable funding for basic research, it is able to employ a large army of research assistants to facilitate the research mission. Because of the generous research funding, the university is able to recruit this assistance from around the world and thus is not so dependent on its own efforts for human resource development.

In the interactive model that tends to characterize the approach of several Asia Pacific settings, the university shares the responsibility for basic research with the other sectors and thus it has relatively less funds to support research and recruit research assistants. However, the universities, especially those in the public sector, have a critical role in the development of human resources for the other sectors. The overall levels of access to higher education are higher than in other regions of the world (NSB, 2004:46), and for those young people pursuing higher education the first- and second-degree training is heavily skewed to science and engineering. For example, in Japan and Korea's public sector approximately 40 percent of all first degrees are in science and engineering. In China, over 50 percent are in these fields. By virtue of this science and engineering (S&E) emphasis, the university systems of Japan, Korea, and Taiwan each graduate a larger proportion of their college age cohort in the natural sciences and engineering than does the United States of America (NSB, 2–35 and 2–39). In terms of the total number of first-degree S&E graduates, China, India, and Japan produce about the same number annually as does the United States of America, with Korea not far behind.

Recent Efforts to Stimulate Creative Research in the Academy and Elsewhere

In the interactive model, universities share many research functions with other sectors. However, especially in recent years, steps have been taken to improve the research environment, especially at the universities. They are given below.

Increased Funding for Research, Including Basic Research

As indicated above, most of the Asia Pacific nations are steadily increasing the resources they are devoting to research and development. Parallel with the overall increase in R&D funds, increasing resources are being channeled to the academic sector.

Science Cities with Universities as the Core

In the mid-1970s following on Russian and American models, Japan launched Tsukuba Science City—research and education center—as its first science city. The new and well-funded Tsukuba University was placed in the center of the city and many government laboratories were moved to this new site. Tax incentives were set up to encourage industrial firms to locate there. Similar developments followed with the relocation of Osaka University and the upgrading of Tohoku University and Kyushu University. Taiwan has established several new science cities, and Singapore has established a Science Park adjacent to the National University of Singapore.

Greater Autonomy for the Universities

In the imitation and innovation phases of higher education development, leading public universities in the Asia Pacific Region tended to be outposts of national policy and subject to extensive regulation by national authorities. With the new push for creativity, the pervasive public regulations including line-item budgets have come to be perceived as obstacles. To erase the bureaucratic feel of these universities, the Indonesian, Japanese, and Thai governments have sought to make universities autonomous statutory authorities with full authority over their resources and operations. These initiatives are being carefully followed by other nations in the region.

Ranking Universities, and/or Ranking Academic Units

With the shift to greater university autonomy, Asia Pacific governments have begun the search for new criteria on which to base public allocations to universities. One possibility is to rank universities and to distribute funds through block grants adjusted by ranking (and other criteria such as total number of students or faculty). China several years ago spoke of focusing central funding on the top 100 universities. In 2001, Minister Aoyama of Japan spoke of focusing funding on the top 25 Japanese universities. In fact, no government has actually implemented these proposals. However, a related principle has been to rank the component units of the many universities in a system and use these unit rankings for preferential funding. Over the past several years, Japan has experimented along these lines with its "Centres of Excellence" Programme.

Peer Review of Research Proposals

In the state-regulated university it was customary to allocate research funds on an equal basis to each academic unit regardless of their productivity or potential. A "new" approach is to require those units and individual professors who desire research funds to prepare a research proposal for anonymous review by a committee of peers. This approach is presumed to elicit more careful development of research programs and to channel funds to those researchers most likely to realize innovative results.

Increased Support of Large- and Medium-Scale Projects of Longer Duration

When research funds were limited, there was a tendency to annually distribute small allocations across the university system. As units could expect to get the same modest amount year after year, this approach did facilitate multiyear research agendas. In keeping with the modest funding, these agendas tended to focus on small problems. But in recent years, R&D policy-makers have come to understand that *big research breakthroughs require big efforts*. Thus, in several of the Asia Pacific systems new funding opportunities are emerging, which encourage large ambitious multiyear projects. In some instances, these are awarded to individuals or groups who work in the conventional academic units. Parallel to these conventional awards, many new and generously funded research institutes are also being established.

Trial Periods for Prospective Researchers

In many Asia Pacific systems, universities were inclined to recruit new staff from among the top students of their recent graduating classes and, in keeping with the spirit of "civil service" appointments, to offer these new employees the equivalent of lifetime tenure. While this personnel policy guaranteed the loyalty of new recruits, it did not always result in the best choices. As many candles "burned out" as continued to "shine brightly." Recognizing the weight of deadwood, many systems—or particular universities within the respective systems—have introduced a trial period for initial appointments.

Efforts to Reclaim "Drained Brains"

Asia Pacific universities "lose" many graduates to the research and development entities of the United States of America and Western Europe (NSB, 2004:2–31). The quality of first-degree training in Asia Pacific universities, especially in the sciences and engineering at the top-ranking universities is quite high. Thus, graduates from these institutions tend to be successful when they apply for graduate education in the West. And many who complete graduate education in the West tend to stay on for post-doctoral and other employment opportunities. China and India are numerically the largest suppliers of foreign talent to the "knowledge industries" of the West, though not an inconsiderable number of young "knowledge workers" migrate from other Asia Pacific countries such as Japan, Korea, Malaysia, and Singapore. But in recent years as the research conditions in the Asia Pacific Region improve, this trend may be changing. There is evidence that more Asia Pacific students are electing to stay home for graduate studies and postdoctoral opportunities. After 2 decades of steady growth in the number of young Chinese people seeking overseas graduate education, their numbers appear to be leveling off since 2001.

Opening the Doors to Foreign Talent

Additionally, Asia Pacific universities are experiencing greater success in recruiting foreign students for their graduate school and postgraduate fellowship opportunities. For example, in Japan in 2001, foreign students made up 8 percent of all Japanese graduate student enrollments in engineering, 10 percent in the natural sciences, and 20 percent in the social sciences (NSB, 2004:2–38). Asia Pacific universities, especially those in the smaller countries that have limited indigenous pools of knowledge workers, are increasing their efforts to attract established professionals from other countries. Most Japanese and Korean universities now have numerous positions available for overseas visiting professors and researchers, and in Singapore higher education institutions advertise internationally for virtually every academic opening. According to a recent study, Japan in 1999 attracted 240,936 high skilled immigrants, an increase of 75 percent over the 1992 figure (Fuess, Jr., 2001). Singapore has been able to attract many outstanding researchers to its laboratories including, recently, a noted biochemist who is a Nobel Laureate.

Asian Science and Technology Is Gaining International Prominence

The Asia Pacific Region's new commitment to research and development is beginning to show results. The most obvious indications are in the application of science and technology for commercial purposes:

1. Asian countries, most notably Japan and Korea, have steadily increased their numbers of domestic patents over the past 2 decades as well as their applications for patents in foreign markets.
2. Asian countries, especially Japan, Korea, and China, have shifted substantial proportions of their industrial production toward high-tech products.

Currently, Korea reports that a higher proportion of its industrial production is in high-tech areas than is the case for the United States of America.
3. Asian nations are also beginning to increase their share of high-tech production in the service industries, a market formally monopolized by the United States of America.
4. Finally, over the past 2 decades several "Other" Asia Pacific nations (China, Malaysia, Singapore, and Taiwan) have been expanding their share of the global market for high-tech products. This combination of countries was supplying less than 8 percent of global high-tech exports in 1980 compared to 30 percent for the United States of America. By 2001, Other Asia's share had increased to 27 percent and the USA share had dropped to 18 percent. During this period, Japan's share dropped from 12 to 10 percent.

Table 2.2 Science and engineering articles, by region and country/economy: 1988–2001

Region and country/Economy	1988	1990	1995	2000	2001
Worldwide	466,419	508,795	580,809	632,781	649,795
OECD	386,267	422,129	487,111	520,349	532,756
North America	199,937	215,389	229,320	222,044	226,704
Canada	21,391	22,792	24,532	22,873	22,626
Mexico	884	1,038	1,901	2,950	3,209
USA	177,662	191,559	202,887	196,221	200,870
Western Europe	143,882	159,898	199,688	225,696	229,173
Austria	2,241	2,690	3,477	4,259	4,526
Belgium	3,586	4,103	5,260	5,739	5,984
Croatia	N.A.	N.A.	563	704	710
Cyprus	15	17	41	60	74
Denmark	3,445	3,716	4,408	4,929	4,988
Finland	2,789	3,071	4,134	4,878	5,098
France	21,409	22,937	29,309	30,960	31,317
Germany	29,292	32,295	38,100	43,440	43,623
Greece	1,239	1,397	2,068	2,892	3,329
Iceland	69	89	158	154	174
Ireland	790	902	1,210	1,596	1,665
Italy	11,229	13,062	17,904	21,038	22,313
Macedonia	0	0	34	49	74
Netherlands	8,581	10,176	12,330	12,466	12,602
Norway	2,192	2,426	2,953	3,195	3,252
Portugal	429	587	989	1,813	2,142
Slovenia	N.A.	N.A.	443	901	876
Spain	5,432	6,837	11,343	14,776	15,570
Sweden	7,573	8,172	9,284	9,815	10,314
Switzerland	5,316	5,901	7,361	8,454	8,107
Turkey	507	750	1,713	3,482	4,098
United Kingdom	36,509	39,069	45,993	49,485	47,660
Yugoslavia	1,211	1,641	507	513	547
All Others	28	58	104	99	129
Asia	51,765	59,282	78,055	104,544	113,575
Bangladesh	95	116	170	160	177
China	4,619	6,285	9,261	18,142	20,978
India	8,882	9,200	9,591	10,047	11,076
Indonesia	59	104	133	165	207

Continued

Table 2.2 Continued

Region and country/Economy	1988	1990	1995	2000	2001
Japan	34,435	38,570	47,603	55,413	57,420
Malaysia	208	233	373	470	494
Pakistan	235	257	339	277	282
Philippines	127	157	151	177	158
Singapore	410	572	1,184	2,301	2,603
South Korea	771	1,170	3,806	9,386	11,037
Sri Lanka	107	106	82	104	76
Taiwan	1,414	2,119	4,846	7,008	8,082
Thailand	287	282	338	655	727
Viet Nam	52	60	102	144	158
All Others	64	50	76	96	101
Eastern Europe/Central Asia	41,597	42,836	36,390	35,844	33,686
Armenia	N.A.	N.A.	182	167	152
Azerbaijan	N.A.	N.A.	157	89	68
Belarus	N.A.	N.A.	728	576	528
Bulgaria	1,089	1,216	963	887	784
Czech Republic	2,746	3,079	1,993	2,458	2,622
Estonia	N.A.	N.A.	240	344	339
Georgia	N.A.	N.A.	148	141	110
Hungary	1,714	1,722	1,826	2,292	2,479
Kazakhstan	N.A.	N.A.	173	113	116
Latvia	N.A.	N.A.	154	159	157
Lithuania	N.A.	N.A.	179	262	272
Moldova	N.A.	N.A.	140	96	77
Poland	4,030	3,999	4,535	5,342	5,686
Romania	393	377	648	956	997
Russia	N.A.	N.A.	19,974	18,271	15,846
Slovakia	N.A.	N.A.	1,137	1,007	955
Ukraine	N.A.	N.A.	2,856	2,365	2,256
USSR	31,625	32,443	N.A.	N.A.	N.A.
Uzbekistan	N.A.	N.A.	295	273	204
All Others	N.A.	N.A.	63	45	39
Near East/North Africa	7,893	8,226	9,627	11,092	11,777
Algeria	66	98	151	204	225
Egypt	1,130	1,254	1,359	1,376	1,548
Iran	86	94	271	825	995
Israel	4,916	4,968	5,921	6,314	6,487
Jordan	161	176	153	242	240
Kuwait	304	368	166	243	257
Lebanon	54	29	56	139	202
Morocco	113	97	237	471	469
Oman	13	27	53	99	96
Saudi Arabia	569	644	781	595	580
Tunisia	96	104	147	278	344
United Arab Emirates	23	33	122	144	159
All Others	362	333	210	161	174
Pacific	12,054	12,962	15,922	17,791	17,743
Australia	9,896	10,664	13,387	14,700	14,788
New Zealand	2,075	2,227	2,466	3,037	2,903
All Others	83	70	69	55	53
Central/South America	4,748	5,848	7,646	11,797	13,147
Argentina	1,423	1,627	1,969	2,792	2,930
Brazil	1,766	2,374	3,471	6,195	7,205

Continued

Table 2.2 Continued

Region and country/Economy	1988	1990	1995	2000	2001
Chile	682	830	899	1,100	1,203
Colombia	86	122	167	320	324
Costa Rica	55	54	70	82	92
Cuba	67	108	166	282	299
Peru	68	77	68	77	93
Uruguay	42	57	98	158	155
Venezuela	292	314	430	509	535
All Others	268	286	307	281	310
Sub-Saharan Africa	4,544	4,355	4,161	3,973	3,990
Cameroon	35	46	73	76	75
Ethiopia	71	70	99	90	93
Ghana	37	40	65	95	90
Kenya	291	255	310	237	230
Nigeria	886	815	464	428	332
Senegal	72	83	77	73	62
South Africa	2,523	2,406	2,364	2,237	2,327
Tanzania	64	69	92	100	87
Uganda	21	29	49	78	91
Zimbabwe	116	131	109	104	113
All Others	425	412	459	457	490

n.a. = not applicable.
Organisation for Economic Co-operation and Development (OECD).

Notes: Article counts are from a set of journals classified and covered by the Institute for Scientific Information's Science Citation and Social Sciences Citation Indexes. Article counts are based on fractional assignments; for example, an article with two authors from different countries is counted as one-half of an article for each country. Countries with article output of less than 0.01 percent of world output in 2001 are grouped in all '*Others*'. Germany's output includes articles from the former East Germany before 1992. China's output includes articles from the Hong Kong economy before 2000. Czech Republic's output includes articles from the former Czechoslovakia before 1996. Article output from the former USSR is included. Details may not add to totals because of rounding.

Sources: Institute for Scientific Information, Science Citation Index and Social Sciences Citation Index; CHI Research, Inc.; and National Science Foundation, Division of Science Resources Statistics, special tabulations. Science & Engineering Indicators – 2004.

Asia Pacific knowledge products, it is often said, are based on foreign technology, but as noted above Asia in recent years has an impressive record in the indigenous development of patents. Japan currently *generates twice as much in revenue from the sale of its patents to foreign entities as it spends on the acquisition of foreign technology*, and the balance sheet for Korea and Taiwan are about equal.

Related to the emerging strength of the Asia Pacific Region in knowledge products is the parallel emergence of a more active and creative academy. One illustration of this new creativity is the increasing prominence of articles written by Asia Pacific scholars in internationally refereed journals. Focusing on articles in the science and engineering fields, both Japan and Other Asia countries have experienced rapid gains in their number of referred articles over the past 15 years, a doubling in the case of Japan and a quadrupling in the case of Other Asia. By way of comparison, the volume of articles written by USA researchers has been stable over this 15-year period and the volume written by Western European scholars has increased by about 65 percent. As a result in 2001 Japanese scholars alone were publishing 13 percent of the world's total and Other Asia an additional 8 percent (see details by country in table 2.2). While the Asia Pacific Region total of 21 percent is less than the USA share of 30 percent, the Asia Pacific proportion has steadily gained in recent years and shows every sign of maintaining that trajectory. While growth in Japan and Korea may slow down, other countries in the region are likely to surge forward.

References and Works Consulted

Altbach, P. G., and T. Umakoshi, eds. (2004) *Asian Universities: Historical Perspectives and Contemporary Challenges*. Baltimore: John Hopkins University Press.

Bartholomew, J. R. (1989) *The Formation of Science in Japan*. New Haven: Yale University Press.

Bush, V. (1945) *Science: The Endless Frontier*. Washington, USA: Government Printing Office, March 1945.

Cummings, W. K. (1990) The Culture of Effective Science: Japan and the United States. *Minerva*, 28(4), Winter: 426–445.

Fuess, S., Jr. (2001) "Highly Skilled Workers and Japan: Is There International Mobility?" Workshop paper presented at Institute for the Study of Labour, Bonn, Germany, August 2.

Hicks, F. B. et al. (2001) *US-Japan Dialogue On the Role of Science and Technology Into the New Millennium*. New York: New York Academy of Sciences.

Johnson, C. (1982) *MITI and the Japanese Miracle*. Stanford: Stanford University Press.

Kimura, S. (1995) *Japan's Science Edge*. Lanham, MD: University Press of America.

Kodama, F. (1991) *Analyzing Japanese High Technologies: The Techno-Paradigm Shift*. London: Pinter Publishers.

Low, M., S. Nakayama, and H. Yoshioka (1999) *Science, Technology and Society in Contemporary Japan*. Cambridge: Cambridge University Press.

Mack, G., and G. Postiglione, eds. (1997) *Handbook of Asian Higher Education*. New York: Garland.

Marginson, S. (2004) "National and Global Competition in Higher Education: Towards a Synthesis (Theoretical Reflections)." ASHE Annual Meetings, Kansas City, Missouri, November 22–24, 2004.

Nakayama, S. (1984) *Academic and Scientific Traditions in China, Japan, and the West*. Tokyo: University of Tokyo Press.

—— (1991) *Science, Technology and Society in Post-War Japan*. London: Kegan Paul International.

NSB (National Science Board) (2004) *Science and Engineering Indicators 2004*. Washington: USA Government Printing Office.

Needham, J., and Wang Ling (1956) *History of Scientific Thought*. Vol. 2 of a six-volume series Science and Civilization In China. Cambridge: Cambridge University Press.

Sarji, A., ed. (1993) *Malaysia's Vision 2020: Understanding the Concept, Implications and Challenges*. Kuala Lumpur: Pelanduk Publications.

Shanghai Jiao Tong University Institute of Higher Education (SJTUIHE) (2003) Academic Ranking of World Universities—2003. http://ed.sjtu.edu.cn/ranking.htm (accessed July 31, 2005).

Stallings, B. (1990) "The Role of Foreign Capital in Economic Development." In Gefeffi and Wyman, eds., *Manufacturing Miracles—Paths of Industrialization in Latin America and East Asia*. Princeton, NJ: Princeton University Press.

World Bank (WB) (1992) *The East Asian Economic Miracle*. New York: Oxford University Press.

CHAPTER THREE

RESEARCH POLICY AND THE CHANGING ROLE OF THE STATE IN THE ASIA PACIFIC REGION

Grant Harman

Introduction

This chapter explores recent changes in national and institutional research policy for higher education in the Asia Pacific Region. It particularly concentrates on the role of the state in university R&D (research and development), public funding of university research, priority setting for research, university research links with industry, and commercialization of university research outputs. While the main emphasis is on the Asia Pacific Region, the discussion is set in a broader international context, with greater emphasis being put on national policies rather than institutional policies. Due to a lack of detailed English language documentation and reliable statistics for many countries of the region, the discussion concentrates mainly on those countries with well-developed innovation systems, where information and statistics are readily available.

The term "research policy" is used in this chapter to refer to *guidelines* and *decisions* expressed in directives, regulations, or laws with regard to the funding and regulation of research activities whereas the term "research management" refers to *implementation* of research policy, including *determination* of strategic directions, *allocation* of resources and roles, and *monitoring* and *evaluation* of performance. Often in public and academic discussions the terms "research policy" and "science policy" are used interchangeably, even though in its limited sense the term "science policy" does not include the social sciences, humanities, and creative arts (Nowotny et al., 2001).

Academics sometimes find it difficult to come to grips with use of terms such as research policy and research management. Understandably, they see university research as activities undertaken by individual academics or groups of academics, who work with a large degree of independence. While generally national and institutional research policies have no intention of diminishing academic independence and creativity more than necessary, governments and other research sponsors increasingly are concerned to ensure that scarce resources are employed *effectively and efficiently*.

While there is an extensive literature on what constitutes research and how this differs across disciplines and various institutional settings (Whiston and Geiger, 1992; Haden and Brink, 1992), in essence research can be defined as critical and creative investigations undertaken on a systematic and rigorous basis, with the aim of extending knowledge or solving particular practical or theoretical problems. Extension of knowledge can be aimed at the the following: (i) discovery of previously unknown phenomena; (ii) development of explanatory theory and its application to new situations; (iii) work that provides significant contributions to particular disciplines; (iv) tackling of problems of social and economic significance; and (v) producing original works of intellectual merit.

Since research activity varies considerably in terms of its disciplinary orientations, objectives, methodologies employed, and end products and their use, policy-makers and sometimes academics, themselves, find it convenient to make various distinctions, such as between "basic" and "applied" research, and between "curiosity-driven" and "problem-driven" research. Many countries now use the Organisation for Economic Co-operation and Development (OECD) categorization for statistical collections and sometimes policy discussions on *pure basic research* (experimental and theoretical work undertaken to acquire knowledge without looking for long-term benefits), *strategic basic research* (experimental and theoretical work undertaken to acquire knowledge in the expectation of useful discoveries), *applied research* (original work undertaken to acquire knowledge with a specific application in view), and *experimental development* (systematic work, using existing knowledge gained from research or practical experience, directed to producing new materials, products, or devices) (OECD, 2002). While various categorization has considerable utility in monitoring research performance and in discussions of research policy issues, it has been subject to considerable criticism, especially as traditional boundaries are becoming increasingly blurred in many new multipartner and transdisciplinary research centers.

Universities are key elements in national innovation and science systems, especially in developed countries. They carry out extensive research activities, train future researchers and other skilled personnel, and generate and communicate new knowledge. University research adds to the overall stock of scientific knowledge from which industrial research draws, while academic laboratories are a source of advanced instrumentation often accessed by industry. Universities undertake research activities for a variety of reasons, but significant reasons are strong academic commitments to the value of research and scholarship that are highlighted in university charters and mission statements and are integral parts of academic and disciplinary cultures. But in addition, universities engage in research because of the status and recognition it attracts, its value in supporting teaching efforts particularly at advanced levels, and its role as a vehicle for providing service to the wider society.

With the increasing recent emphasis on international business competitiveness, the production, application, and use of new knowledge generated by universities have taken on increasing importance. Traditionally high quality university research was regarded as work that breaks new ground or is innovative; is systematic and rigorous, with appropriate in-depth analysis or synthesis; and leads to publication or other forms of dissemination so that the findings are open to peer scrutiny and assessment, and are available for the benefit of the wider community. However, with the increasing commercialization of university research and new partnerships between universities and industry, some traditional academic commitments and values are being challenged, such as open dissemination of research results and sharing of research materials among scientists.

Apart from universities, other institutions of importance in science and innovation systems are public sector research institutions (PRIs) and research laboratories operated by business firms. However, the balance of activity among these three different types of research-performing bodies varies considerably among different countries and within countries over time.

The research functions of universities have changed to a major extent over the past 2 centuries. The classical European university concept of research-based teaching (established by Wilhelm von Humboldt of the University of Berlin in 1810) continues to be influential today, with the idea of the modern research university having been developed in the United States of America in the second half of the nineteenth century. In the early part of the twentieth century, the "research university idea" spread widely to many other industrial countries, with a strong emphasis being placed on the role of the university focusing primarily on basic research and research training, with some commitment to *applied research* but little to *developmental research*. The mission and fundamental values of the university at this stage were only moderately tied to the economy and employment of graduates.

From World War II onward, however, demands made on scientific research for reasons of national defense and economic and social development brought universities more directly into contact with research users, leading to ongoing efforts to reform and redirect university research. While the classical university idea retains considerable appeal within academia, increasingly modern universities are being forced to accept a wider research role and to become more directly involved with business, industry, and government (OECD, 1998). Combined with this has been the rapid growth in student enrollments and moves toward more strongly market-driven approaches, with students, firms, governments, and other "customers" placing increasing pressures on university directions and priorities.

Universities in the Asia Pacific Region are responding to these changes in the context of a new emphasis on the knowledge revolution whose focus is on the ability to create access to and use of knowledge as fundamental determinants of global competitiveness. According to the World Bank (WB) "Knowledge for Development Project," key determinants of national success in this knowledge revolution include increased codification of knowledge and development of new technologies; closer links between industry and the national science base, increased importance of education and up-skilling of the labor force; the importance of investment in "intangibles" such as R&D, education, and software; and the desirability of innovation and productivity increases being more important in competitiveness than gross domestic product (GDP) growth (Dahlman, 2002). This rapidly changing context provides particular challenges for countries that need to develop strategies in order to use new and existing knowledge to improve performance in traditional sectors, exploit opportunities for "leapfrogging," and develop new sectors. Even the smallest and poorest countries need to address these challenges in order to support advanced level university training and ensure some level of capacity to access new internationally used technologies.

Higher Education and Research in the Asia Pacific Region

The Asia Pacific Region is a vast collection of states with over 3 billion people, and containing almost 60 percent of the world's population but only one-third of the world's higher education enrollments and only about 30 percent of the world's wealth. It includes 2 countries each with a population in excess of 1 billion people, as well as many small nation states. Countries of the region differ greatly in ethnicity, social characteristics, and the extent of their recent economic development, with striking differences between rich and poor countries, and between rural and urban areas. Sharp contrasts also exist between exceedingly wealthy nations and poor countries, and between nations that have for many years operated market economies and newly independent countries of Central Asia (Harman, 1998).

The diversity of the region is clearly reflected in its higher education systems. The region includes not only some of the largest higher education systems in the world but also microsystems that cater to small numbers of students. Some higher education systems are amongst the strongest and best resourced internationally, while others struggle to find sufficient resources even to provide the most basic elements of higher education provision. Systems such as those of the Indian subcontinent are still largely public systems, whereas in Indonesia, Japan, Korea, the Philippines, and Thailand, a high proportion of students are now enrolled in private higher education. Table 3.1 illustrates the diversity in enrollment size, staff numbers and the production

Table 3.1 Size of selected Asia and Pacific higher education systems as measured by total enrollments, teaching staff, and graduates

Nation	Total enrollements	Teaching staff	Total graduates
Australia	845,132	—	151,862
Bangladesh	878,537	47,137	—
Bhutan	1,837	164	—
Cambodia	25,416	2,124	—
China	12,143,723	679,888	1,948,080
Hong Kong SAR	128,052	10,063	—
India	9,404,460	399,023	—
Indonesia	3,017,887	217,403	476,971
Japan	3,972,468	477,161	1,068,878
Lao	16,621	1,372	2,924
Malaysia	549,205	20,473	—
Mongolia	84,970	6,575	14,868
Myanmar	553,456	10,522	—
Nepal	103,290	—	—
New Zealand	171,962	11,252	42,791
Philippines	2,432,002	93,956	351,078
Republic of Korea	3,003,498	144,185	519,719
Vietnam	749,914	32,977	121,292

Source: Global Education Digest, 2003.

of graduates. By far the largest highest education systems are those of China, India, Japan, and Korea, while some of the smallest are those of Bhutan and Mongolia.

Higher education in the future information age is likely to play a greater role in preparing the region's labor force than in the past. Some analysts have argued that the economic returns to higher education are rising with a shift to science-based industries and services, and according to the Asian Development Bank (ADB, 2003:28) this is being borne out by empirical data. Many countries in the region also dramatically expanded higher education enrollments in the 1980s and 1990s. China, Hong Kong, Korea, the Philippines, Singapore, and Thailand all now have relatively high levels of age cohorts enrolled in tertiary education.

National capacity in research within higher education systems and overall capacity in R&D within the region vary considerably. Table 3.2 provides UNESCO data on the number of researchers per million of population and overall R&D expenditure as a percentage of gross national product (GNP). By far the strongest countries within the Asia Pacific Region in terms of researchers per million inhabitants are Australia, Japan, Korea, and Singapore while in terms of R&D expenditure in relation to GNP the two strongest economies are Japan and Korea, and then at a lower level Australia and Singapore. In some countries, particularly China and India, rapid transformation driven by substantial public and private investment is taking place. It should be noted that the figure provided in this table for Japan is considerably higher than figures that appear in some OECD publications—since OECD uses more conservative estimates of R&D expenditure in Japanese universities than does the Japanese Statistics Bureau (Stenberg, 2004:25).

Leaders in innovation and higher education research within the Asia Pacific Region perform well within the international context. In fact, total expenditure on R&D activities in China, Japan, and Korea combined in 2001 has been estimated to have been approximately equal to that of the whole European Union (EU) (Stenberg, 2004:19).

Table 3.2 Researchers and R&D expenditure in selected Asia and Pacific nations

Nation	Year	Researchers per million inhabitants	R&D Expenditure as % of GNP
Australia	2000	3,439	1.53
Bangladesh	1995	52	0.03
China	2001	584	0.30
Hong Kong SAR	1998	—	0.44
India	1996	157	0.78
Indonesia	1998	182	0.07
Japan	2001	5,321	3.09
Republic of Korea	2001	2,880	2.96
Malaysia	1996	93	0.24
New Zealand	2001	—	1.03
Pakistan	1996	72	0.92
Singapore	2001	4,052	2.11
Sri Lanka	1996	191	0.19
Thailand	1997	374	0.10

Source: UNESCO, 2003.

Japan accounted for 16.7 percent of total OECD R&D expenditure in 2001, while Korea accounted for 3.4 percent. Over the period 1995–2001, Japan invested US$103.8 billion in domestic R&D and Korea US$22.3 billion. While in Japan and Korea business is the main source of funding for R&D, in both Australia and New Zealand business funding is less important than government funding (OECD, 2003a:18–21). In Australia and New Zealand, universities are the major source of basic research, but they coexist with public sector institutions devoted to particular sectors of national interest such as defense, energy, and agriculture. In East Asian countries, which were formerly oriented toward technical applications and the assimilation of foreign technology, university research remained relatively modest for many years, owing to lack of financial support, overregulation, and heavy teaching responsibilities of academic staff. However, in recent years Japan, in particular, has boosted efforts in basic research (OECD, 1998:22).

Other comparative information is available from the World Bank Knowledge for Development Program that has made estimates of the performance of a large number of Asian countries on a number of indicators, including economic incentive regimes, education, information and communications technology capacity, and innovation. These figures demonstrate impressive achievements in a number of Asia Pacific countries and substantial improvements in others. On economic incentive regimes, Hong Kong and Singapore stand out, while on education Korea and Japan are at the forefront, with big improvements having been made in Indonesia, Malaysia, and Vietnam. On information and communications technology capacity, Hong Kong, Japan, Korea, Singapore, and Taiwan are the leaders, with substantial recent improvements having been made in China, India, and Vietnam as well as in Malaysia and the Philippines. In terms of innovation Japan, Korea, and Singapore do well at the higher end of the spectrum. Malaysia performs well particularly in relation to manufacturing, while China and India have a critical mass of researchers who give them advantages over others. Significant improvements in innovation have been made in Indonesia, the Philippines, and Vietnam (Dahlman, 2002).

Role of the State in Research and Innovation

Governments today are generally playing increasingly important roles in funding, stimulating, and directing research activity. In doing so, they use a variety of "policy instruments" or strategies to achieve particular objectives, including allocation of block grants and specific purpose funds to institutions, research centers, and individual researchers and research groups; establishment of major research centers and institutes; investment in major research equipment, provision of economic incentives and disincentives (including subsidies, pricing structures, and taxation concessions or charges); and regulation (such as legislation relating to Intellectual Property [IP]) and the provision of information. In addition, ministers and officials increasingly use persuasion and advocacy. Some of the least understood instruments for encouraging R&D and research commercialization are taxation concessions such as the Australian 125 percent deduction for business R&D expenditure.

Governments within the Asia Pacific Region currently are faced with various pressures and new challenges related to their role in research.

First, governments are being forced to respond to demands from a more diverse set of stakeholders. Traditionally research was seen as having two main stakeholders; (i) the research community; and (ii) those who fund research. Under these arrangements, research universities and researchers largely determined research agendas, while governments saw their responsibilities essentially in maintaining capacity in knowledge creation that could benefit society and provide spillovers to the economic sector. In recent years, however, a larger group of stakeholders are demanding involvement in establishing research priorities and deciding on financial allocations, while the business community itself is carrying out more research and being more involved in supporting particular types of university research.

Second, important changes are taking place in the research enterprise and efforts to capture research benefits to meet social and economic needs. Especially in scientific and technological research, costs are rapidly increasing with the use of highly sophisticated and expensive equipment and other infrastructure support. There also is a general shift away from an almost exclusive emphasis on disciplinary research, which some see as hindering fruitful synergies across fields, toward more multidisciplinary research that is more directly responsive to societal needs and is carried on with more interaction among different research performers. This trend has been described by Gibbons et al. (1994) as the shift from Mode 1 to Mode 2 research. Mode 1 research is generated within traditional disciplinary and cognitive context, while Mode 2 research emphasizes the importance of the application of knowledge, transdisciplinarity, and being more socially accountable and reflexive. While there is ongoing debate about the extent to which Mode 2 research is actually replacing Mode 1 research, it is clear that more multidisciplinary research is occurring at frontiers of scientific research, with many of the most exciting breakthroughs taking place at interfaces between traditional disciplines. This trend is often closely associated with proliferation in the channels for bringing new science and technology (S&T) to the market place, with licensing and spin-offs combining more effectively with role of venture capital, new IP legislation, and new forms of labor mobility.

Third, countries face major challenges in ensuring the long-term sustainability of their research enterprises, particularly in maintaining breadth and diversity in research capacity and ensuring a supply of highly trained human resources. Supply of highly trained scientists is closely related to capacity in PhD training, the ability to secure PhD training abroad or attract foreign-trained scientists, and success in addressing problems of brain drain to other countries.

Fourth, in many countries there are increasing pressures for public accountability, while the ideas of new public sector management (PSM) are being increasingly applied to the research sector. As in other areas of public spending, pressures are increasing for greater efficiency in research investment, while stakeholders are making new or increased demands related to research directions and emphasis in areas such as health, environment, and energy.

Governments have responded to these new challenges in different ways. According to the recent OECD (2003b) report entitled "Governance of Public Research," based on an extensive survey, OECD member countries have responded by improving stakeholder involvement in priority setting with a wider range of stakeholders being involved; restructuring research funding by redefining responsibilities, combining agencies, or developing new mechanisms of coordination; reviewing and renewing R&D funding mechanisms with a strong emphasis on use of competitive arrangements based on performance and merit; undertaking major funding initiatives to strengthen infrastructure support; encouraging enhanced partnerships among universities, public sector research organizations, and private firms; and reform and restructuring of public sector research institutions (PRIs). These trends clearly have been evident in OECD member countries within the region.

In Japan, major administrative reform of the science system took place at the beginning of 2001, with establishment of a central coordinating body for science and technology policy. Increased autonomy was given to national research institutions and national universities, while the ministry responsible for education and science was merged into one ministry with the agency responsible for implementing research (OECD, 2003b). Earlier a new basic law for science and technology had been enacted in 1995 requiring the Japanese government to develop science and technology (S&T) plans for periods of 5 years at a time. In Japan there has been strong pressure favoring radical changes in the system of financing and performing research and innovation, with an attraction to overseas models particularly that of the United States of America (Stenberg, 2004).

In Australia, the Australian Research Council (ARC) and the National Health and Medical Research Council (NHMRC) were given increased independence with new legislation and their own secretariats. All universities were required to submit annual research and research training reports, while a new performance-based funding system was introduced for research and research training, and national research priorities were identified (OECD, 2003b). More recently, a commissioned report has recommended closer collaboration between universities and major publicly funded research agencies and establishment of a single Strategic Research Council with an overall coordinating role (Department of Education, Science and Training, 2004). In Korea, a long-term strategic initiative for science and technology development and a 5-year plan for S&T were established.

Public Funding of Research and Research and Development (R&D)

Public funding of university research and R&D is one of the major instruments used by governments to steer science systems and to capture more effectively economic and social benefits. Many countries have embarked on reforms of their funding systems in response to new demands and opportunities, enhancing their strategic planning capacity and paying more attention to the social and economic environment and to the evolving patterns of relationships between stakeholders. Overall the volume of R&D funding has increased, although public funding is generally increasing at a lesser rate than private funding.

Traditionally in industrialized countries a high proportion of university research was financed by governments as a "public good," but in the 1990s such funding declined with the result that universities were increasingly forced to seek new sources of support. Meanwhile, government funding increased for mission-oriented and contract-based research, more dependent on output and performance criteria. This forced universities to perform more short-term and market-oriented research.

According to an OECD report (2003b), almost all OECD countries have increased R&D funding over recent years, although generally such increases have achieved little more than keep pace with expansion of economies. As a share of GDP, funding for R&D in universities and other public research institutes remained flat at about 0.61 percent between 1981 and 2000 for OECD countries generally, although there were some major variations between countries. Nearly all countries in the OECD study reported their intention to increase funding in future, but generally increases are expected to be mainly in priority areas and in new programs such as centers of excellence where funding is on the basis of competitive grants.

Different types of funding mechanisms are in wide use, particularly institutional or block grants, project funding, and special programs funding, but in each case there is increased use of competitive mechanisms and funding allocations being based on performance. *Institutional* or *block grant funding* takes different forms, although in most countries traditionally it was based on student enrollments or number of research units (or chairs such as in Japan). Generally, such funding comes without strings attached, although Korea is an exception. However, more recently clear trends are to separate "institutional funding for research" from "institutional funding for teaching," and for "allocations for mission-oriented funding" on a competitive basis to be on an increase while long-term general institutional funding declines. Australia, Hong Kong, New Zealand, and the Philippines all use separate streams of institutional funding for research, with allocations based on quality and/or performance. While Australia has used simple performance indicators (external research grants, higher degree completions, and publication outputs), Hong Kong and New Zealand have opted for modified versions of the United Kingdom's Research Assessment Exercise (RAE) based on assessments of research quality conducted every 4 years by some 70 different panels of experts. While throughout the 1990s the Australian "research quantum" scheme allocated about 5 percent of total operating grant funding on the basis of performance indicators, the Higher Education Funding Council For England (HEFCE) allocated some 20 percent of total government funding on the basis of RAE assessments, resulting in leading research universities gaining larger amounts from their research allocations compared to their teaching allocations (Harman, 2000). Since the early 1990s, the University Grants Committee (UGC) Hong Kong has used a modified and less expensive form of the UK RAE to allocate institutional funding, while New Zealand has introduced a performance-based research allocation system aimed to identify and reward researcher excellence with the hope of increasing the average quality of research (Investing in Excellence, 2004; MOE, 2002). Allocations will be based on the quality of researchers (60 percent), research degree completions (25 percent), and external research income (15 percent).

As far as *project funding* is concerned, allocations are made on the basis of applications submitted in response to notifications or calls for tenders and are evaluated usually by peer review processes. Project funding is similar to business funding of R&D in that it tends to be contract-based, with specific objectives and milestones. A decade ago the Australian Research Council (ARC) expanded its range of project funding to include grants for projects with joint industry support.

Special programs are becoming increasingly common. These generally are linked to priority areas, and funding is allocated on a competitive basis, often for centers of excellence, or special research centers involving universities and other partners. Centers of excellence have been established in many Asia Pacific countries including Australia, Japan, and New Zealand. Japan launched a new university resource allocation prioritization program in 2002 called the 21st Century Centre of Excellence (COE) Programme with the aim of promoting research units of world-class excellence in selected fields. The fields supported in 2002 were life science, chemistry and materials science, information, electrical and electronics, humanities, and interdisciplinary subjects. Each research unit selected is being allocated resources around JPY 100–500 million for 5 years. In November 2002, some 113 research units at 50 institutions were selected out of 464 applications from 163 institutions. Australia has programs supporting special research centers, key centers of teaching and research, cooperative research centers (multisite centers jointly funded by government and industry), and a small group of megacenters in strategic areas such as biotechnology and information communications technology (ICT). In 2001, New Zealand established a Centres of Research Excellence Fund to support world-class centers expected to be involved in both research and knowledge transfer activities (MOE, 2001).

In a number of advanced countries, an important trend is for increased R&D to be financed and performed by business. However, in Japan, business support for higher education and public research institutions has increased only slightly but it is still relatively small, while in Korea business funding for higher education research has decreased but this reduction has been compensated by increased funding from the government, with an increase over the past 2 decades of about 100 percent. Other funding for university research comes from institution-owned resources, endowments, and patent licensing fees. In Japan and Korea, 5 percent or more of research funding comes from other sources.

Across advanced regional economies governments are increasingly linking evaluation and assessment with funding allocations. Detailed assessments sometimes are made prior to new initiatives, while ongoing assessment of performance is increasingly common. Traditionally evaluation procedures were mainly based on the use of peer review of project applications, but governments now are using in-depth reviews and performance indicators such as total external funding attracted, and numbers of publications, patents, start-ups, awards, and prizes.

Priority Setting

Priority setting by governments and universities is a process of strategic choice with the aim of increasing returns on investment in research. In this process some fields of research or particular research centers or research projects are selected over others to receive preferential funding. Both government and university priorities are being reflected in research funding decisions and reforms of funding mechanisms.

Priority setting is a complex and difficult political process involving many participants and taking different forms. Important distinctions can be made between different forms of priority setting—for example, between thematic priorities (e.g., such as improving health care) as opposed to structural priorities (e.g., different funding instruments), between disciplines (e.g., sciences as opposed to humanities) as opposed to priorities between different forms of research (e.g., basic versus more applied research), and between relatively short-term as opposed to medium- or longer-term plans.

Priority setting has had a number of drivers. Governments are increasingly aware of the direct relevance of knowledge gained through science to economic growth and social well-being and so look for higher returns on their research investments. This is often accompanied by stronger accountability demands and the application of competitive mechanisms and other new public sector management (PSM) ideas (as in Australia and New Zealand). In some cases, priority setting is driven by reductions in government budgets but more commonly priority setting is favored to make decisions about how increased research budget allocations should be spent. Still again, many governments face strong pressures to provide larger allocations to particular areas, such as health or environmental studies, and so need to "free up" funding from other areas. In Korea, identification of research priorities is directly linked to selecting engines of future economic growth.

Priority setting is difficult because of competing pressures, existing rigidities particularly with highly decentralized funding in some countries, and the need to respond to new opportunities and societal needs. Shifting priorities within constrained budgets is particularly difficult, and so often only any increased component of budgets are allocated by priorities. In the United Kingdom, for example, only annual increases in the science budget are allocated to priority areas as identified collectively by the various "Research Councils" and through the "Foresight Exercise."

Governments in the Asia Pacific Region use a variety of priority setting mechanisms, including "national science and technology plans," "advisory bodies," and "foresight processes and public consultation," while universities tend to depend on "strategic plans," "research management plans," and particular competitive "funding mechanisms." Since 1970, Japan has been conducting periodic technology forecasting exercises using the Delphi method, while Korea uses foresight and the results are implicitly integrated into national priorities by experts who are involved in evaluation and prebudget review. In many countries governments have made deliberate attempts to centralize and coordinate priority setting. In Hong Kong priority areas have been established by an Areas of Excellence Sub-Committee of the University Grants Committee, following recommendations of the Sutherland Committee of Review that indicated that Hong Kong needed world-class institutions with distinct areas of excellence in order to retain its leading economic position in the development of China and the Pacific Rim. To date, eight areas of excellence (including information technology, economics and business strategy, molecular neuroscience, and Chinese medicine) have been selected for a period of 5 years in 3 rounds of funding with a total of HK$320 million being allocated (University Grants Committee, 2004). Frequently, national priority setting processes have broader participation to include scientific experts together with business and community representative in the interest of increasing transparency as well as in response to the genuine requirement to better respond to societal needs. In the Philippines choice of priority research areas for the "National Higher Education Agenda" were guided by the principles of multidisciplinarity, policy orientation, and possible impact.

In some countries a "top-down" approach is dominant such in as Hungary, Japan, Norway, and where the central government adopts explicit strategies, policies, or plans that specify priority areas for research. In these countries, plus others such as Denmark, Germany, Korea, and the Netherlands, there is some form of central advisory body that gives recommendations about priorities. As already noted, Japan launched a new university resource allocation prioritization scheme called "The 21st Century COE Programme in 2002" with the aim of promoting research units of world class in selected fields (OECD, 2003b:98). At the other end of the spectrum, is the "bottom-up" decentralized approach such as in Canada, Sweden, and the United States of America where advisory bodies relate to different government agencies in priority setting.

In the past Australia depended on a sectoral and pluralist approach to priority setting, with priorities being set within major policy domains, often resulting in strong competition between research and operations in health, education, or energy. However, the Commonwealth government's innovation plan released in January 2001, "Backing Australia's Ability" (Howard, 2001), flagged the need for an emphasis on research in which Australia enjoys or wants to achieve a competitive advantage. A significant shift in priority setting was announced by the Minister for Education, Science and Training in January 2002 when four research priority areas were announced for the "Australian Research Council's 2003 Funding" under the "National Competitive Grants Programme." More recently in 2002 the Australian government began a new process to identify national research priorities that would influence the agenda of all major Commonwealth government research funding agencies. This involved extensive consultation and development of a short list of priorities by an expert committee from more than 180 submissions. From the list of priorities in December 2002 the government identified four thematic priorities: (i) environmentally sustainable Australia; (ii) promoting and maintaining good health; (iii) frontier technologies for building and transforming Australian industries; and (iv) safeguarding Australia. Public research bodies are required annually to put forward plans to government on how they propose to implement the priorities.

Traditionally within universities researchers set their own priorities within their own projects, while at department and faculty levels decisions on allocations were often made using relatively informal processes. But now many universities are being forced to take priority setting and selective funding more seriously, often leading to considerable tensions.

Research Links with Industry

As a result of institutional initiatives and government encouragement and financial support, in many countries universities have established much closer and more effective links with other research providers and stakeholders, particularly public sector research institutions (PRIs), industry laboratories, business firms, and government agencies. These links take a variety of different forms including joint research centers and research appointments, shared use of facilities, industry funding of university research, and consultancy arrangements between universities and research users. In the United Kingdom, for example, a rapid increase in university-industry collaboration since the 1980s has led to a variety of different partnership arrangements with many positive outcomes including an impressive increase in the number of joint scientific publications. By the late 1990s, joint university-industry papers amounted to about half of all industrial scientific output (Calvert and Patel, 2002). These new arrangements have increasingly broken down traditional arrangements whereby in modern economies universities and PRIs are viewed as being responsible for basic scientific and precommercial research, while industrial firms perform the bulk of applied research and product development (Hall, 2004).

On the whole, these developments have worked well for the mutual benefit of the various partners and have contributed to successful innovation efforts. University research links with industry provide universities with substantial research support, consulting opportunities, support for postgraduate students, opportunities for graduate employment, and opportunities for academics to gain insights into new developments within industry; while industry benefits through access to university expertise and facilities, access to university intellectual property, and supply of well-trained graduates. Admittedly considerable tensions are sometimes generated and even scientists themselves acknowledge that there are risks involved.

Particularly important partnerships in a number of countries are new research centers with multiuniversity partners as well as partners from PRIs, government departments, and business firms. An example in the Asia Pacific Region is the Cooperative Research Centres (CRC) Australia, a program that has resulted in establishment of some 70 multisite centers. Funding is provided by the Australian government as well as from partners. Some CRC are set up using a company structure, while others are unincorporated using the legal basis of one or more partners.

While governments, universities, and the researchers involved in partnerships are generally supportive of university-industry partnerships, critics allege that such partnerships threaten traditional academic values, lead to distortions in the balance between basic and applied research, and tend to corrupt academics with commercial values to the extent that some academics neglect their responsibilities in teaching and research. It is also alleged that industry contracts lead researchers to withhold scientific information from colleagues and delay publication and thus adversely affect the free flow of scientific information.

Various evaluative studies have investigated various aspects of the impact of the new industry links on universities and academic work. The United States of America points to the dangers in these new relationships particularly the impact on academic work and values, forcing scientists to abandon the traditional cooperative mode of research (Dickson, 1984; Kenney, 1986). In their multinational study of academic capitalism Slaughter and Leslie (1997) reported that while senior academics often respond positively to opportunities to attract funds from industry, many junior academics are confused and ambivalent, having "difficulty conceiving of careers for themselves which merged academic capitalism and conventional academic endeavour" (Slaughter and Leslie, 1997:173).

Other scholars (i.e., Etzkowitz and Peters, 1991), however, provide evidence to support the claim that many academic researchers increasingly accept the concept that profit generated from research need not corrupt and conclude that to date there has not been any great effect on academic behavior with regard to direct industry funding of academic research. Particularly important in addressing criticisms of the new commercialism have been the detailed studies of researcher behavior by

Blumenthal and colleagues. One study (Blumenthal et al., 1986) that reported on a survey of 1,200 academic researchers in 40 major universities in the United States of America, in the area of biotechnology found that researchers with industrial support publish at higher rates, patent more frequently, participate in more administrative and professional activities, and earn more than colleagues without such support. At the same time, researchers with industry funds are much more likely than other biotechnology researchers to report that their research has resulted in trade secrets and that "commercial considerations" have influenced their choice of research projects.

Some USA findings have been largely confirmed by Australian studies and evaluations. Various performance indicators point to the considerable success of efforts by the Australian government to enhance university-industry links while officially sponsored evaluations and reviews point to a high level of overall success for particular programs. Studies of science and technology academics in leading Australian universities also show that researchers with industry funding tend to be more senior and more likely to hold national competitive grants than colleagues without industry funding. Industry-funded academics also have better publication records, spend longer hours at work each week, and more time on postgraduate teaching, administration, committee work, and interaction with colleagues (Harman, 1999). Another study of science and technology academics in five leading research-intensive universities revealed that an estimated 40 percent of regular academic staff enjoyed industry funding with about 60 percent of these having attracted funding in excess of AU$250,000 over the past 3 years (Harman, 2002). About 40 percent of respondents with industry funding reported having conducted research where the results are the property of a sponsor and cannot be published for a period without consent. Half of these admitted having delayed publication for more than 6 months but safeguarding the researcher's self-interest was as a common a motive for delaying publication as was protecting the property of a sponsor.

University Research Commercialization and Technology Transfer

Since the early 1980s first in the United States of America, and more recently in many other developed countries, governments and research-intensive universities have been putting much more effort into enhancing capacity in research commercialization and in the transfer of university-generated inventions and discoveries to the commercial sector. These developments have been driven partly by the wish of universities to generate additional income, but universities also have become increasingly involved in commercialization activities to enhance relationships with firms and to generate political support by demonstrating the positive outcomes of public investment in research. Governments, on the other hand, seek to capture the benefits of university research to facilitate economic and social development and to generate wealth.

The terms "research commercialisation" and "technology transfer" often are used synonymously, although strictly speaking there are important differences in their precise meanings. Research commercialization refers to the process of turning scientific discoveries and inventions into marketable products and services. Generally university research outputs are commercialized by licensing patents to companies or by the creation of "spin-out" companies that usually depend on assignment of university intellectual property (IP) for their initiation. In the scholarly literature, the term "technology transfer" has a number of specialist meanings but in essence refers to "the movement of know-how, technical knowledge, or technology from one organization to another" (Bozeman, 2000:629). The most common use of the term is in relation to the transfer of inventions and associated "know-how" from research organizations (especially universities and PRIs) to research users.

Research commercialization and technology transfer is based on IP rights of which patents, industrial designs, copyrights, and trademarks are the most important. IP rights reward investment in R&D by granting ownership to inventors, their employers, those who funded the research, or some combination. Over the past 2 decades, governments and universities have become increasingly aware of the value of IP and various strategies that can be employed to derive commercial and public benefit.

Licensing of inventions and the creation of new companies, of course, are not the only mechanisms of research commercialization employed by universities since both graduates and academics regularly carry knowledge from universities to business firms, while industry accesses university-based knowledge through sponsored research, conferences, and academic journals (Sizer, 2002). However, increasingly licensing and creating companies are seen as key mechanisms of university research commercialization.

In a number of Asia Pacific countries, governments and universities are allocating increased funds to support research commercialization. Frequently, governments have a multiplicity of programs with numerous agencies being involved, raising questions about policy coherence and coordination, and about whether or not large corporations tend to benefit more than SMEs (small- to medium-sized enterprises) and universities. Some countries clearly are doing better than others in terms of measured outputs and economic growth rates, while within countries there are notable examples of particular regional successes. This raises important questions about the effectiveness of different combinations of government and university strategies, about the relative amounts of funding involved, and about how such funding is employed and with what success.

Why some countries are more successful than others in commercialization of university research appears to be dependent on a variety of factors, particularly government financial support and the regulatory framework, incentive systems operating to affect the behavior of universities and researchers, institutional culture, and the legal basis relating to the ownership and commercialization of intellectual property (IP). Important recent contributions have been made by the Swedish economist Magnus Henrekson in combination with two USA-based colleagues, Nathan Rosenberg and Brent Goldfarb (Goldfarb and Henrekson, 2003; Henrekson and Rosenberg, 2001). These scholars argue that the United States of America has been far more successful than Sweden in the commercialization of university research despite Sweden's strong research base. They attribute the different success particularly to different government roles, a stronger incentive structure in the United States of America for both universities and academics to be actively involved in research commercialization, and the legal basis for intellectual property. While Sweden has employed a largely government-led "top-down" approach with an academic environment that discourages academics from actively participating in commercializing their ideas, the USA approach has been strongly "bottom-up," with government IP legislation providing strong incentives for institutional and academic involvement in research commercialization. This has been combined with a highly competitive USA education environment.

According to Henrekson and Rosenberg (2001), the USA bottom-up approach led essentially by major research universities, which, under the Bayh-Dole Act of 1980, had ownership of all IP resulting from federal research grants. In this situation, their argument is that both federal and state governments did relatively little to develop new government agencies or other mechanisms to enhance university capacity in technology transfer. This argument appears to have considerable validity, but it needs some modification in view of recent major investments by numerous USA state governments in expensive research infrastructure (Geiger, 2003).

While a number of Continental European countries appear to follow a Swedish-type model (Gittelman, 2002), in many other countries including the United Kingdom, Canada, Australia, and New Zealand there is a mixed approach, with emphasis being placed on new government support and incentive programs for industry and universities, as well as on strong incentive systems for universities and academics. The role of incentives clearly is of great importance in any theory explaining the growth of science-based entrepreneurship. At the same time, governments clearly can play important roles to support science-based entrepreneurship, from providing incentives to universities such as via the Bayh-Dole Act in the United States of America with regard to IP ownership to providing different forms of subsidies, grants, loan funds, and guides on good practice. In many cases, major government emphasis has concentrated particularly on providing seed funding to assist early development phases of commercialization of inventions that have commercial potential as well as various programs of grants and loans to assist companies and to encourage university-industry collaboration. In the United Kingdom, for example, in combination with the "Wellcome Trust" and the "Gatsby Charitable Foundation," the government established the "University Challenge Fund" to provide seed funds to groups of universities for early-stage R&D (ARC, 2000:14). In a relatively small number of cases, governments have established new specialized commercialization agencies, such as in Sweden, where since 1994 seven broker institutions, called technology bridging foundations, have been established in major university regions (Henrekson and Rosenberg, 2001). Their task has been to mediate commercialization of R&D from universities and researchers to small- and medium-sized enterprises by facilitating patenting processes and matching up researchers with venture capital funding. These foundations have been designed to accept some of the responsibilities that in the United States of America lie with technology licensing offices on university campuses.

Relatively little detailed data has been available on the research commercialization successes of different countries outside of the United Kingdom, the United States of America, Canada, and Australia for which survey data is readily available. For this reason, in 2001 the OECD Committee for Scientific and Technological Policy commissioned a project to collect empirical evidence about patenting and licensing activity in universities and public sector research institutions (PRIs) in OECD countries as well as information on the legal and regulatory frameworks that govern IP. While the data presented in the project report (OECD, 2003b) need to be treated with caution, they give clues about comparative national performance and point to a range of issues that are in need of investigation.

Laws and policies governing the ownership of IP are being revised in a number of countries, generally with a view to encouraging ownership of IP by institutions performing the research. In Japan and Korea, recent reforms in funding regulations have given universities more control over the IP generated by their researchers. These reforms echo the landmark of the USA Bayh-Dole Act of 1980. Changes in IP laws in Korea, for example, have been driven by recognition that a considerable amount of university and PRI research is not being channeled to industry in a timely manner. In Australia, universities are able to claim IP rights since it is a general principle of common law that an employer is entitled to any intellectual property rights created by an employee in the course of his/her employment. Furthermore, both the Australian Research Council (ARC) and the National Health and Medical Research Council (NHMRC) have specifically stated that they do not claim ownership over IP resulting from research they fund (Christie et al., 2003).

A major barrier has been lack of financial incentives for universities, PRIs and researchers, and inability of institutions to take responsibility for the cost of management of IP. This is well illustrated in the case of Korea where in the late 1990s it was recognized that, despite increasing investment in Korea in R&D, the share of patent applications was still surprisingly small. Despite the fact that the public sector accounted for about 27 percent of investment in R&D in the late 1990s, it only accounted for less than 5 percent of patent applications. Furthermore, a 1997 survey by the Intellectual Property Office (IPO), Korea, revealed that only 31 percent of total patents awarded were licensed (Yun, 2003). Legally public universities and PRIs have operated under different patent laws compared to private universities, with IP in public universities being the property of the state. Intellectual property management arrangements were modified by the "Technology Transfer Facilitation Law, 2000" that unified IP management in all public institutions, requiring the establishment of technology transfer offices and sharing of proceeds of license income between inventors and institutions, and by amendment of the Patent Law in 2001 allowing public universities to gain financially from patent licensing (Yun, 2003:240–250). Further legislative changes followed in 2002 that resulted in transfer of ownership of inventions from professors to transfer license organizations set up by universities while more recently the Intellectual Property Office, Korea, has designated 55 universities for special support in IP creation and management (Choi, 2003).

Since it was recognized in Japan that intellectual property issues cross the boundaries of many ministries, in 2002 a Strategic Council on Intellectual Property was established "in order to quickly establish and advance a national strategy for intellectual property" (Stenberg, 2004:17; Motohashi, 2003). This led to enactment of a new "Basic Law on Intellectual Property, 2002" that particularly aimed to encourage the creation of IP in universities and increased international standardization, with particular measures directed at facilitating the establishment of technology licensing offices in universities and supporting the education of specialists in IP law.

Many other countries in the region are reviewing and strengthening intellectual property legislation and management, although in many cases other issues than intellectual property ownership within universities are of central importance. In China, further strengthening of the role of the State Intellectual Property Office (SIPO) has occurred following membership of the World Trade Organization (WTO). This has concentrated particularly on administrative issues and issues related to software piracy. In India recent effort has concentrated particularly on "modernization of patent information services" and "modernization of the trademarks registry."

Expansion of research commercialization activity, and as a direct consequence of legislation, to give intellectual property rights for all or most university research to institutions has stimulated the development of research commercialization offices that are concerned with filing patent applications, entering into licensing agreements with third parties, and being involved in the creation of spin-out companies. These developments have required considerable institutional, financial, and human

Table 3.3 Summary results from OECD survey on patenting and licensing activities in universities and public research institutes

		Patents		Licenses		Start-ups and spin-offs:
		Total patent stock in last year	Number filed in last year	Licences earning income	Gross income Euro	Not created in past year 2002
Australia (2000)	All	—	834	491	99,525	47
	Uni	—	586	—	79,834	32
	PRI	—	248	—	19,691	15
Belgium (2001)	All	506	121	4	240	15
Germany (2001)	PRI	5,404	1,058	1,188	66,368	37
Italy (2000)	All	—	190	8436	—	—
	Uni	—	102	1227	—	—
	PRI	—	88	72	—	9
Japan (2000)	All	682	567	324	1,397	6
Korea (2001)	All	9,391	1,692	132	3,822	56
	Uni	404	244	22	1,032	19
	PRI	8,987	1,448	110	2,790	37
Netherlands	All	991	212	93	11,400	37
(2000)	Uni	394	111	—	—	27
	PRI	597	101	—	—	10
Norway (2001)	PRI	114	43	39	7,700	51
Spain (2001)	All	781	133	136	961	11
Switzerland	All	1,184	175	77	5,650	68
(2001)	Uni	914	132	61	2,800	56
	PRI	270	43	16	2,850	12
USA	All	—	8,294	—	—	—
	Uni	—	6,135	8,670	12,974	390
	PRI	—	2,159	484	52	—
Russia	All	—	171	8	1,375	15

Source: OECD, 2003c.

Notes: Australia: Gross income in US$; Italy: number of patent applications estimates; Netherlands: gross income is an estimate; USA: total number of earning licenses for federal laboratories is underestimated, and income is in US$; Russia: patent applications are estimates.

resources, but in many countries the direct contributions of governments have been limited. However, in a small number of countries, including Japan, governments have provided short-term support to universities to assist in covering the costs of patenting and commercializing inventions (OECD, 2003c:13).

In a number of countries including Australia, China, Japan, Korea, and New Zealand universities have expanded existing research management offices, created new in-house research commercialization offices, or established specialized offices with an arms-length relationship using company structures. Denmark, Germany, Korea, and the United Kingdom are experimenting with regional- or sector-based technology transfer offices to manage technology transfer activities for groups of universities and public research institutes. Potential economies of scale might be realized by spreading fixed costs over a greater number of institutions and exploiting the advantages of portfolio diversification, but these models may find difficulty in developing close working relationships with researchers. With the alternative model, a recent development is to transfer local technology specialists in university faculties, responsible to both deans and a central university technology transfer office.

The size of patent portfolios and stock of currently active patents varies considerably between countries, as does income and number of "start-ups" and "spin-outs." Summary data from the OECD survey are shown in table 3.3. Korea clearly is a major player internationally in patenting and licensing.

While much of the recent Korean increase in patenting has been attributed to expansion of biotechnology, patenting is also significant in other fields including health, information technology, food, and energy. Patenting outcomes usually reflect a country's R&D and industrial specialization. In Korea, for example, where international communications technology (ICT) is important in business-value-added production, over 70 percent of universities reported having filed patents in ICT and electronics.

Conclusion

Countries across the Asia Pacific Region are facing unprecedented changes in their higher education systems as they come to grips with fundamental economic and social change, and the impact of globalization and increasing international economic competition. These changes are impacting significantly on university research policy, with clear trends toward some *redefinition* of the role of universities and *redirection* of their research activities. Leading higher education systems in the region are following similar developments to those found generally in OECD countries, particularly

- changing roles for the state in research policy and innovation;
- establishment of new mechanisms for allocating public funding of research;
- experiments in priority setting and research concentration;
- enhanced university-industry partnerships; and
- more serious efforts to capture research outputs in order to create jobs and produce economic and social benefits.

On the one hand many other countries of the region are following similar trends, particularly China, India, Malaysia, and Thailand, although statistical data and detailed information is less easy to access. On the other hand, many of the poorer countries find great difficulty in meeting enrollment pressures let alone allocating significant sums to support research.

Although the discussion has concentrated mainly on national policy particularly in science and technology disciplines that are related to public funding and priority setting, industry business links and research commercialization, it is important to recognize that significant changes are affecting research in a wide range of different academic disciplines and are providing difficult challenges for vice-chancellors and presidents, especially in relation to developing strategic priorities and implementing mechanisms for research funding selectivity.

References and works consulted

ADB (Asian Development Bank) (2003) "Key Indicators 2003: Education for Global Participation." Manila.

ARC (Australian Research Council) (2000) "Research in the National Interest: Commercializing University Research in Australia." Report. Canberra: Australian Government Publishing Service.

Blumenthal, D. et al. (1986) "University-Industry Research Relationships in Biotechnology: Implications for the University." *Science*, 232:1361–1366.

Bozeman, B. (2000) "Technology Transfer and Public Policy: A Review of Research and Theory." *Research Policy*, 29:627–655.

Calvert, J., and P. Patel (2002) *University-Industry Research Collaboration in the UK*. Brighton: SPRU (Science and Technology Policy Research Unit), University of Sussex.

Choi, J. (2003) "Creation, Management and Use of IP—An Integrated and Proactive IP Policy." Paper prepared for WIPO Asia Pacific Regional Seminar on Intellectual Property Strategy for Economic Development, Kuala Lumpur, December 9–11.

Christie, A. F., S. D'Aloisio, K. L. Gaita, M. J. Howlett, and E. M. Webster (2003) *Analysis of the Legal Framework for Patent Ownership in Publicly-Funded Research Institutions*. Canberra: Commonwealth Department of Education, Science and Training.

Dahlman, C. (2002) *Knowledge for Development: Assessment Framework and Benchmarking*. Washington: Knowledge for Development, World Bank.

Department of Education, Science and Training (2004) "Review of Closer Collaboration between Universities and Major Publicly-Funded Agencies." Canberra.

Dickson, S. (1984) *The New Politics Of Science*. New York: Basic Books.

Etzkowitz, H., and L. Peters (1991) "Profiting from Knowledge: Organisational Innovations and the Evolution of Academic Norms." *Minerva*, XXIX:133–166.

Geiger, R. (2003) "Beyond Technology Transfer: New State Policies for Economic Development for USA Universities." Paper presented at 16th CHER Annual Conference, Porto, October.

Gibbons, M. et al. (1994) *The New Production of Knowledge: The Dynamics of Science and Research in Contemporary Societies*. London: Sage.

Gittleman, M. (2002) "The Institutional Origins of National Innovation Performance: Careers, Organizations, and Patents in Biotechnology in the USA and France." New York, New York University (unpublished manuscript).

Global Education Digest (2003) *Comparing Education Statistics across the World (2003)*. Montreal: UNESCO Institute for Statistics.

Goldfarb, B., and M. Henrekson (2003) "Bottom-Up Versus Top-Down Policies Towards the Commercialization of University Intellectual Property." *Research Policy*, 32(4):639–658.

Haden, C. R., and J. R. Brink, eds. (1992) *Innovative Models for University Research*. Amsterdam: North Holland.

Hall, B. H. (2004) "University-Industry Research Partnerships in the USA." Conference Paper, Kansa, Kansai, February.
Harman, G. (1998) "Institutional Challenges and Response Strategies in National Strategies for Regional Co-Operation in the 21st Century." In *Proceedings of the Regional Conference on Higher Education*, Tokyo July 8–10, 1997. Bangkok: UNESCO Principal Regional Office for Asia and the Pacific.
—— (1999) "Science and Technology Academics and University-Industry Research Links in Australian Universities." *Higher Education*, 38(1):83–103.
—— (2000) "Allocating Research Infrastructure Grants in Post-Binary Higher Education Systems; British and Australian Approaches." *Journal of Higher Education Policy and Management*, 22(2):111–126.
—— (2002) "Australian University-Industry Research Links: Researcher Involvement, Outputs, Personal Benefits And 'Withholding' Behaviour." *Prometheus*, 20(2):143–158.
Henrekson, M., and N. Rosenberg (2001) "Designing Efficient Institutions for Science-Based Entrepreneurship: Lessons from the USA and Sweden." *Journal of Technology Transfer*, 26(3):207–231.
Howard, J. (2001) "Backing Australia's Ability: Real Results, Real Jobs." Report. Canberra: Commonwealth of Australia.
Investing in Excellence (2004) *The Report of the Performance-Based Research Fund Working Group*. Auckland: Ministry of Education.
Kenney, M. (1986) *Biotechnology: The University Industrial Complex*. New Haven: Yale University Press.
MOE (Ministry of Education) (2001) "The Centres of Research Excellence Fund." Report. Ministry of Education, Auckland.
—— (2002) "Development of the Performance-Based Research Fund." Auckland.
Motohashi, K. (2003) "Japan's Patent System and Business Innovation: Reassessing Pro-Patent Policies." Institute of Innovation Research, Hitotsubashi University and RIETI.
Nowotny, H., P. Scott, and M. Gibbons (2001) *Re-Thinking Science: Knowledge and the Public in an Age of Uncertainty*. Cambridge: Polity Press.
OECD (Organisation for Economic Co-operation and Development) (1998) *University Research in Transition*. Paris: OECD.
—— (2002) *Frascati Manual: Proposed Standard Practice for Surveys on Research and Experimental Development*. Paris: OECD.
—— (2003a) *OECD Science, Technology and Industry (STI) Scoreboard*. Paris: OECD.
—— (2003b) *Governance of Public Research: Towards Better Practices*. Paris: OECD.
—— (2003c) *Turning Science Into Business: Patenting and Licensing at Public Research Organization*. Paris: OECD, 15.
Sizer, J. (2002) "Research and the Knowledge Age." *Tertiary Education and Management*, 7:227–242.
Slaughter, S., and L. Leslie (1997) *Academic Capitalism: Politics, Policies, and the Entrepreneurial University*. Baltimore: John Hopkins University Press.
Stenberg, L. (2004) *Government Research and Innovation Policies in Japan*. Ostersund: Swedish Institute for Growth Policy Studies.
UNESCO (2003) *Selected R&D Indicators*. Paris: UNESCO.
UGC (University Grants Committee) (2004) *Areas of Excellence*. Hong Kong: University Grants Committee.
Whiston, T. G., and R. L. Geiger (1992) *Research and Higher Education: The UK and the USA*. Buckingham: Society for Research into Higher Education and Open University Press.
Yun, M. (2003) *Regulatory Regime Governing Management of Intellectual Property of the Korean Public Research Organizations: Focus on the Biotechnology Sector in Turning Science Into Business: Patenting and Licensing at Public Research Organisation*. Paris: OECD.

CHAPTER FOUR

THE CHANGING LANDSCAPE OF HIGHER EDUCATION RESEARCH POLICY IN AUSTRALIA

V. Lynn Meek

Introduction

Over most of the twentieth century, higher education has been shaped by the norms of science, democracy, and the need for an educated citizenry, cultural preservation, and trained bureaucratic elite. But with the advent of the so-called knowledge economy/knowledge society, higher education has been recognized by politicians, industrialists, and some academics as well, not only as a creator and transmitter of knowledge but also as "a major agent of economic growth: the knowledge factory, as it were, at the centre of the knowledge economy" (*The Economist*, 1997). According to Scott (1997), "[H]igher education systems are no longer simply 'knowledge' institutions, reproducing the intellectual and human capital required by industrial society; they are becoming key instruments of the reflexivity which defines the post-industrial [and postmodern] condition."

The challenges currently faced by the contemporary university are not a result of increasing lack of relevance to society but are due to its very success. As the knowledge society continues to develop, the university is faced with a growing number of competitors in both research and training. Yet, what is not in doubt is the continuing importance and centrality of the university as knowledge is increasingly brought within market and political exchanges. The question, therefore, is not if the university will survive but rather in what form and playing what role(s)? Society imposes new roles, pressures, and demands on higher education while simultaneously expecting the preservation of key traditional functions (Clark, 1998; Neave, 2002). Higher education institutions in turn help shape the very society that generates these new and traditional expectations. The modern university subsequently operates in a climate of what Barnett (2000) refers to as a "world of super-complexity." The university simultaneously helps generate this "super-complexity" and is asked to assist in resolving the uncertainties it generates.

As part of a wider agenda of public sector reform, new approaches to higher education steering and coordination increasingly shift from government control to forms of market-like coordination. This trend toward marketization and privatization of public sector higher education has been well established over the last decade or more and is clearly visible both in the language of policy documents (students as customers and clients [Jackson, 2002], knowledge as a product or commodity, price and quality relations, etc.) and in their implementation: the introduction of tuition fees, performance-based funding, and conditional contracting (Hayden, 2003). The introduction of these market-like mechanisms makes the environment in which higher education institutions must operate all the more fluid and turbulent.

However, while acknowledging the increasing importance of the market in defining the role and purpose of higher education, it needs to be recognized that the marketplace is only one of several forces shaping the structure and character of higher education institutions and systems. It would be a mistake to reduce all the forces bringing about change in higher education to market relations. Analysis of current trends must both incorporate the importance of market steering of higher education while accounting for other social, political, and cultural forces that help shape the sector and individual institutions. As Scott (2003:212) puts it,

> The lesson drawn by many political [and university] leaders was that the way forward for higher education was to abandon collectivist public-service public-sector policies and practices and embrace the "market"; universities must seize the opportunity to become the leading organizations in the burgeoning global knowledge economy. Not to seize this opportunity was to risk marginalization—even, eventually, extinction. The discussion of the impact of globalization in higher education continues to be dominated by this neo-liberal orthodoxy, but it is this orthodoxy [better, ideology] that must be challenged if universities are successfully to embrace the "world," in all its problematical diversity, rather than simply the global marketplace.

Where, in the past, universities had a sense of shared intellectual purpose (at least to a degree), bolstered by the security of centralized funding and control, at present they are confronted by a much more complex, fluid, and varied environment that articulates different, and sometimes conflicting, demands, thus creating new and complex realities. Consequently, new distributions of authority emerge, new (accountability) relationships arise amongst constituents inside and outside the university, and a new dynamic within policy fields develops. Clearly, Australian higher education, and the development of research within it, is caught up in such a dynamic. Australia more so than many (if not most) countries has gone down the market path with mixed results. Thus, the nation's higher education system may serve as an important case study for higher education policy development elsewhere.

This chapter commences with a brief sketch of the background to the Australian higher education system and the role of research and development (R&D) within it. This is followed by a summary of government reforms of higher education and a profile of the present state of the sector. The chapter then turns to an examination of the various policy reviews and their recommendations that have helped shape the structure and character of higher education research over the past 10 years (1996–2005) or so. The chapter concludes with an analysis of the outcomes of various policy implementations and the issues that they raise.

Background to the Australian Higher Education System—Role of R&D

Australia is a federation of six states and two territories. An exceptional feature of the higher education sector is that the states have legislative control of higher education institutions, whilst financial responsibility (since 1974) rests with the Commonwealth. The nation's higher education sector consists of 37 public universities, some of which are quite large with enrollments in excess of 45,000 students, 2 small private universities, and a number of both public and private small specialist institutions.

Whereas in terms of landmass Australia is the sixth largest country in the world—approximately the same size as the continental United States—it has a population only slightly larger than the Netherlands. Most of the nation's population of some 20 million people (0.3 percent of world population) is highly urbanized. "The country's economy is 1.9 per cent of the Gross Domestic Product (GDP) of the OECD, and accounts for about 1 per cent of world trade," Department of Education, Science and Training (DEST, 2003a:3). Historically, the nation's wealth was based on primary products—mineral and agricultural. But in recent decades there has been a deliberate attempt by government and industry to switch the basis of the Australian economy from primary products to knowledge—to create what one prime minister termed in the 1980s as the Clever Country. While in the early 1970s, about 21 percent of Australia's GDP was based on manufacturing and 5.4 percent on agriculture, presently those figures are 12 percent and 3.6 percent respectively. As the chief economist of one of the country's largest banks put it, "Australia's economic growth will increasingly be linked to the mortarboard not the sheep's back." (*The Sydney Morning Herald*, 2004).

Australia has a well-developed but comparatively small science base, with the majority of its R&D effort concentrated in the public sector. Taking into account the size of the nation, Australia's contribution to world science is impressive, particularly with respect to medical and health disciplines and biological sciences and astronomy. Based on 2002 data, Australia

1. [C]ontributed 2.88 per cent of the world's output of research publications (including in the sciences, social sciences and humanities), up from 2.3 per cent in 1988;
2. was ranked ninth out of 21 countries behind Canada, France, Germany, Italy, Japan, Spain, the United Kingdom (UK) and the United States of America (USA) in the total number of research publications and ahead of countries such as Korea, the Netherlands, Sweden and Switzerland;
3. was ranked eighth out of 21 countries in the number of research publications on a *per capita* basis, ahead of Canada, France, Germany, Japan and the USA and behind Denmark, Finland, Israel, Netherlands Switzerland and the UK (DEST 2003a:6).

For a number of historical and geographical reasons, the funding of Australian R&D is more highly dependent upon the public purse than is the case in most other developed countries. In 2000, government-financed expenditure on R&D was 0.71 percent of GDP, compared to an OECD average of 0.64 percent (DEST, 2003a:18).

In contrast, business expenditure on research and development (BERD) is low compared to other OECD (Organisation for Economic Co-operation and Development) countries. This is largely due to the fact that most of the large multinational corporations in Australia have their headquarters elsewhere and conduct little of their R&D in this country (Gallagher, 2000).

Unlike the United States of America and United Kingdom, there are very few private foundations for Australians to look to for research support (Wills, 2001), and there is nowhere the level of endowment support that some of the major USA universities enjoy. While there has been some resent recovery, BERD as a percentage of GDP reached a peak in 1996 of nearly 0.9 percent, after which it declined sharply to about 0.65 percent in 2000, largely due to the impact of the government's change to the R&D tax concession from 150 percent to 125 percent in the last half of the 1990s.

Australia's BERD as a percentage of GDP in 2001 was less than half of the Organisation for Economic Cooperation and Development (OECD) average of 1.62 percent, and "in 2000 Australia ranked sixteenth in the OECD in the share of gross expenditure on R&D undertaken by business (47 per cent) compared to the OECD average of just under 70 per cent" (DEST, 2003b:25).

The remaining gross expenditure on total R&D was 23 percent from state and federal government, 27 percent from higher education, and 3 percent from the private nonprofit sector (Australian Vice-Chancellors' Committee [AVCC—the committee of Australia's university presidents], 2003a:14).

Australia also lags behind many other OECD countries in terms of gross domestic expenditure on R&D (GERD) as a proportion of gross domestic product. Australia's total expenditure is 1.53 percent of GDP compared to an OECD average of 2.25 percent. There have been calls from such bodies as the AVCC that Australia should increase its investment in research to 2 percent of GDP by 2010 and 3 percent by 2020.

Table 4.1 provides an overview of Australian R&D performance relative to that of other OECD countries in terms of key indicators: Gross Domestic Expenditure on R&D (GERD), business expenditure on research and development (BERD), Expenditure on R&D in the Higher Education sector, government intramural expenditure on R&D (GOVERD), government-financed GERD, and industry-financed GERD. In terms of these indicators, Australia performs above the OECD average with respect to public sector investment in R&D, but, as already indicated, below average overall.

The relatively low level of investment in R&D from the private sector has meant that government has had to play a leading role in funding Australian science and innovation. The federal government channels its support for R&D through a variety of schemes and organizations, the two major being the Commonwealth Scientific and Industrial Research Organisation (CSIRO) and the nations universities, the former receiving about AU$612 million direct from government and the latter AU$6,652 million. Of course, Commonwealth support for universities is for teaching as well as research. About 27 percent of GERD is performed by the higher education sector (see table 4.2), which is a fairly large proportion relative to many other OECD countries.

Table 4.1 Overview of international R&D performance by OECD countries, 2000

	GERD (million current GDP—AU$)	GERD/ GDP	BERD/ GDP	HERD/ GDP	GOVERD/ GDP	Govt-financed GERD/ GDP	Industry-financed GERD/ GDP
Total OECD	604575.0	2.25	1.56	0.38	0.23	0.64	1.44
USA	265,179.6	2.72	2.04	0.38	0.18	0.71	1.88
Japan	98,389.3	2.98	2.11	0.43	0.29	0.58	2.16
Germany	53,568.9	2.49	1.75	0.40	0.34	0.79	1.64
France	32,873.2	2.18	1.37	0.41	0.38	0.84	1.15
UK	27,184.0	1.85	1.21	0.38	0.22	0.53	0.91
Korea	18,939.6	2.65	1.96	0.30	0.35	0.64	1.92
Canada	16,193.4	1.87	1.09	0.55	0.22	0.58	0.79
Italy	15,482.8	1.07	0.53	0.33	0.20	—	—
Sweden	8,879.4	3.96	3.03	0.81	0.12	0.90	2.77
Netherlands	8,440.2	1.94	1.11	0.57	0.25	0.70	0.97
Australia	7,743.3	1.53	0.72	0.41	0.35	0.71	0.70
Spain	7,568.2	0.94	.050	0.28	0.15	0.36	0.47
Switzerland	5,600.8	2.63	1.96	0.60	0.03	0.61	1.82
Belgium	4,944.7	1.96	1.40	0.47	0.06	0.45	1.30
Finland	4,457.0	3.40	2.41	0.61	0.36	0.89	2.39
Mexico	3,505.0	0.43	0.11	0.11	0.19	0.26	0.10
Denmark	3,204.1	2.19	1.42	0.43	0.32	0.68	1.09
Turkey	2,685.3	0.64	0.21	0.39	0.04	0.32	0.28
Poland	2,583.3	0.70	0.25	0.22	0.23	0.44	0.23
Norway	2,430.3	1.64	0.95	0.45	0.25	0.67	0.83
Czech Republic	1,892.5	1.33	0.80	0.19	0.34	0.59	0.68
Portugal	1,358.9	0.79	0.22	0.29	0.19	0.51	0.22
Ireland	1,235.6	1.15	0.83	0.23	0.09	0.26	0.76
Greece	1,123.0	0.67	0.19	0.33	0.15	0.33	0.16
Hungary	998.6	0.80	0.36	0.19	0.21	0.40	0.30
New Zealand	760.7	1.03	0.31	0.35	0.37	0.52	0.35
Iceland	219.4	2.77	1.56	0.45	0.71	—	—

Source: DEST, 2003c:12.

Also, a greater proportion of Australia's R&D workforce is located in higher education than is the case for most OECD countries (see table 4.3).

Table 4.4 gives a rough idea of university expenditure on R&D by source of funds. One interesting aspect of this table is the small but steady increase in R&D expenditure from state and local government sources. As mentioned above, in 1974 almost total funding for higher education was assumed by the Commonwealth, and since then the funding and policy influence of state governments on higher education has been on the wane. But in recent years, some states have been targeting university funding particularly in the biotechnology fields in the belief that such investment will strengthen the local economy—a clear commitment to the notion of the knowledge economy. Several states have established science and innovation councils under such banners as Queensland's Smart State; Victoria's Science, Technology and Innovation Initiative; New South Wale's Bio-First Strategy; and Western Australia's (WA) Innovate Policy (Meek, 2002).

Table 4.2 GERD performed by sector and OECD country, 2000

Country	Percentage share of GERD			
	Business	Government	Higher education	Private non-profit
Total OECD	69.6	10.4	17.1	2.9
Sweden	76.4	3.1	20.4	0.1
USA	75.2	6.8	13.9	4.1
Korea	74.0	13.3	11.3	1.4
Switzerland	73.9	1.3	22.9	1.9
Ireland	71.8	8.1	20.1	—
Japan	71.0	9.9	14.5	4.6
Finland	70.9	10.6	17.8	0.7
Germany	70.3	13.6	16.1	—
Slovak Republic	65.8	24.7	9.5	—
UK	65.6	12.2	20.8	1.4
Denmark	64.9	14.5	19.4	1.2
France	62.5	17.3	18.8	1.4
Czech Republic	60.0	25.3	14.2	0.5
Canada	58.3	11.9	29.5	0.3
Norway	57.8	15.0	27.2	—
Netherlands	57.1	13.0	29.2	0.7
Iceland	56.4	25.5	16.2	1.9
Spain	53.7	15.8	29.6	0.9
Italy	50.1	18.9	31.0	—
Australia	47.1	23.1	27.1	2.7
Hungary	44.3	26.1	24.0	5.6
Poland	36.1	32.2	31.5	0.2
New Zealand	29.7	36.0	34.3	—
Greece	28.5	21.7	49.5	0.3
Portugal	28.2	24.3	37.2	10.3
Mexico	25.5	45.0	26.3	3.2

Source: DEST, 2003c:19.

Table 4.3 Researchers per 10,000 labor force by sector of employment

Country	Business	Government	Higher education
USA	70	4	10
Japan	64	5	26
Sweden	52	6	33
Finland	41	16	40
Germany	34	10	17
Ireland	33	2	15
UK	32	5	17
Canada	31	5	21
Denmark	28	14	20
France	28	9	22
EU	25	7	18
Netherlands	23	10	16
Australia	15	10	41
New Zealand	9	9	26

Source: AVCC, 2003b:10.

Table 4.4 University expenditure on research and experimental development by source of funds, 1988–2000 (percentages)

Source of funds	1988 (%)	1990 (%)	1992 (%)	1995 (%)	1996 (%)	1998 (%)	2000 (%)
Commonwealth government:	—	—	—	—	—	—	—
General university funds	—	—	—	66.1	65.4	63.7	62.9
Commonwealth schemes	—	—	—	16.9	16.3	16.6	17.4
Other Commonwealth government	—	—	—	6.6	7.0	7.4	6.4
Total Commonwealth	91.3	88.4	91.1	89.6	88.7	87.7	86.8
State and local government	1.5	2.5	2.1	2.2	2.2	2.7	3.2
Business enterprise	2.6	2.2	2.5	4.7	5.2	5.2	4.9
Other Australian	3.9	6.2	3.7	2.5	2.8	2.9	3.0
Overseas	0.7	0.7	0.6	1.1	1.1	1.6	2.2

Source: AVCC, 2003a:6.

Table 4.5 Research income by source, 2001 (percentages)

Research income source	Amount (AU$)	Total resource income (%)
Commonwealth competitive grants	475,337,497	40.88
Non-Commonwealth competitive grants	14,364,311	1.24
Total national competitive grants	489,701,808	42.12
Local government	4,691,588	0.40
State government	101,714,402	8.75
Other Commonwealth government	93,601,769	8.05
Total other public sector funding	200,007,759	17.20
Australian contracts	119,014,269	10.24
Australian grants	70,262.074	6.04
Donations, bequests, and foundations	64,926,726	5.58
International funding	137,089,095	11.79
Syndicated R&D	1,622,206	0.14
Total industry and other funding	392,914,370	33.79
Total government grants (excluding CRCs)	689,709,567	59.32
Total financial data (excluding CRCs)	1,082,623,937	93.11
Cooperative Research Centre (CRC) funding		
Commonwealth grants to CRCs	58,245,303	5.01
Nonuniversity participants	12,806,703	1.10
Third party contributions	9,032,234	0.78
Total CRC funding	80,084,240	6.89
Total research income (public universities)	1,162,708,177	100.00

Source: AVCC, 2003a:11.

Table 4.5 gives a more detailed picture in terms of money specifically targeted for research—the government component of full-time staff salaries (25 percent) nominally devoted to research is excluded. As the table illustrates, about two-thirds of university income specifically for research comes from government sources and one-third comes from business and industry.

History of Government Reforms of Higher Education

Throughout the 1970s and into the 1980s, policy-makers and institutional leaders alike became increasingly concerned about the future of Australian higher education. This culminated in a push at the end of the 1980s to make higher education more relevant to national economic needs and priorities. The 1988 White Paper outlining federal government policy initiated a dramatic transformation of Australian higher education that, amongst other things, led to the abolition of the binary distinction between universities and colleges of advanced education (CAEs) and the creation of the unified national system (UNS) in which there is now a much smaller number of significantly larger institutions, all called universities. The reforms also placed the need for selectivity and concentration of research squarely on the agenda. These events are often referred to as the Dawkins' Reforms, in recognition of one of their primary architects, the then federal minister of Employment, Education and Training, the Honorable John Dawkins.

In July 1988 the White Paper on higher education was adopted by the federal government and set in motion a period characterized by the Following: (i) the dismantling of the binary system; (ii) a challenging of the view that teaching and research are inextricably linked; (iii) the emergence of new systems of funding and emphasis for higher education institutions to diversify their funding sources; (iv) a sharper sense of the real importance of research for economic well-being; (v) a growing appreciation that for relatively small countries such as Australia, concentration and selectivity are essentials in any national research policy; and (vi) a much greater emphasis on institutional management (Dawkins, 1988).

The major policy shifts can be summarized as follows:

1. A shift in some of the cost of higher education from the state to the individual: the government lessened its financial commitment through the introduction of mechanisms such as the Higher Education Contribution Scheme (HECS, partial tuition payment through the tax system).
2. Enhanced national and international competition for students and research income is another shift.
3. Greater emphasis is to be laid on accountability for the government dollar.
4. There should be greater deregulation within the higher education sector.
5. Reliance should increase on income gained from sources other than the Commonwealth.
6. Expectation should be clear that higher education contributes to economic prosperity and the knowledge economy.

Diversity, quality, and coordination of the higher education sector were key policy intentions of the White Paper and have continued to be so despite the change of government. The White Paper is quite clear regarding the Unified National System (UNS) not being a uniform system by stressing the following:

1. The new arrangements will promote greater diversity in higher education rather than any artificial equalization of institutional roles.
2. The ultimate goal is a balanced system of high quality institutions, each with its particular areas of strength and specialization but coordinated in such a way as to provide a comprehensive range of higher education offerings.
3. Diversity and quality are paramount; the unified system will not be a uniform system.

However, the sector's responses to these policy initiatives have not necessarily been in accord with the initial intentions. In particular there have been numerous unintended consequences resulting from the changed policy framework—this is especially true in the areas of research management, funding, and training. At the national level degrees of concentration and selectivity have not occurred to the extent expected from the policies. However, a more informal concentration and grouping of research universities has occurred. A relatively new and interesting phenomenon resulting from competition is the creation of alliances and networks of various types such as Universitas 21, Group of 8, the Australian Technology Network, and Innovative Research Universities Australia (Wood and Meek, 2002).

With the change of federal government in March 1996, it became clear that the size of the task to which higher education must adapt had, in fact, substantially increased. The 1996 budget statement from the newly elected Liberal coalition government regarding higher education placed additional pressures and challenges on this sector. Key changes announced in the 1996 budget statement included the following:

1. A reduction of operating grants by 5 percent over 3 years
2. A lowering of the HECS repayment threshold, an increase in level of HECS payments, and the introduction of differential HECS according to course of study
3. No Commonwealth supplementation of academic salary increases
4. An insistence upon return of funds if enrollment targets are not met
5. A phasing out of postgraduate coursework enrollments from Commonwealth-funded load

The funding changes have had a profound and largely negative effect on higher education from which the sector is still reeling. Total public investment in Australian universities peaked in the mid-1990s and then decreased through to 2001. The funding cuts to higher education initiated in 1996 did not really start to bite until the end of the decade. But with the advent of the new millennium, it has been generally recognized that Australian higher education faces a funding crisis (Chubb, 2000 and 2001). Funding of Australian higher education increased during the period 1995–2000 with respect to all sources of revenue. However, direct public funding from the Commonwealth government declined by 11 percent in real terms—Australia being only 1 of 2 OECD countries in which this occurred. And, while total funding increased by 12.5 percent in real terms, total student load increased by 21 percent (Phillips et al., 2002:28). Research funding per se has also suffered.

Australia's investment in research and development has fallen steadily in the past few years, at a time when many of the nation's major competitors and trading partners are increasing their commitment to research and innovation.

In 2000–2001, total national expenditure on research and development was just over AU$10 billion, or around 1.5 percent of GDP—below most of the OECD countries and half the target level set by Canada, the European Union, and others (AVCC, 2003b:13).

Nearly all of the recent reviews and changes to Australian higher education have attempted to address the funding issue in one form or another—with government primarily relying on market mechanisms rather than increased public subsidies to solve the problem. The remainder of this chapter will summarize recent reviews and policy changes to higher education that have had a direct impact on research and research management. But first a few words need to be said about the size and structure of the present higher education system.

Current Profile of Australian Higher Education

In 2004 Australian universities enrolled nearly 1 million students, about 25 percent of whom were full fee-paying international students. The overseas student market is worth more than AU$6 billion annually to Australia and makes it one of the nation's largest export earners. Fees paid directly to higher education institutions from overseas students rose from AU$627 million in 1997 to AU$1,423 billion in 2001. Overseas students contribute about 13 percent to the total higher education budget.

Since the late 1980s, there has been substantial growth in Australian higher education, from about 485,000 students in 1990 to more than double that in 2004. However, in recent years, most of the student growth has been fueled by overseas students. In the period 1995–2001, the number of commencing domestic students increased by 8.6 percent, while the number of commencing overseas students rose by 146 percent; however, the slow growth in domestic student numbers does not indicate a slacking in demand but lack of available places to meet demand (Phillips et al., 2002:8).

Funding of Australian higher education increased during the period 1995–2000 with respect to all sources of revenue (see table 4.6). However, direct public funding from the Commonwealth government declined by 11 percent in real terms. And, while total funding increased by 12.5 percent in real terms, total student load increased by 21 percent (Phillips et al., 2002:28).

The government says itself that it no longer funds but subsidizes higher education. The proportion of the budget going to higher education from the Commonwealth government varies according to whether or not HECS is included as part of the Commonwealth grants. If HECS is counted separately, then less than 40 percent of the revenue for higher education comes directly from the Commonwealth (see table 4.7).

Table 4.6 University revenue by source 1995–2000 (AU$ in billions) (adjusted by CPI to 2000 terms)

Funding Source	1995	1996	1997	1998	1999	2000	(%) change
Commonwealth	4.7	4.9	4.7	4.6	4.4	4.2	−11.0
HECS	1.0	1.0	1.3	1.5	1.7	1.7	68.9
Fees	1.0	1.2	1.3	1.4	1.6	1.7	75.3
State	0.1	0.1	0.1	0.1	0.1	0.1	25.8
Other	1.5	1.5	1.4	1.3	1.3	1.6	7.9
Total	8.3	8.6	8.8	9.0	9.1	9.3	12.5

Source: Phillips et al., 2002:26.

Table 4.7 Higher education institution operating revenue by source, 2003

Source	AU$	(%)
Commonwealth government funding	4,919,513	39.89
State government	506,042	4.10
HECS	1,917,206	15.55
Postgraduate education loan scheme	178,950	1.45
Fees and charges	2,720,720	22.06
Investment income	318,678	2.58
Consultancy and contract research	637,500	5.17
Other income	1,133,217	9.19
Total	12,331,826	100.00

Source: DEST, 2004.

It should be noted that the Higher Education Contribution Schemes' (HECS) revenue is returned to the higher education system *via* the HECS Special Account. The Commonwealth contributes the difference between the repayments received and the total HECS payments required to be made to the sector (the latter being total HECS liability minus upfront payments). For 2002–2003, total student repayments were expected to be AU$848 million. This comprised AU$137 million in voluntary repayments and an estimated AU$710 million repaid via the taxation system. These repayments represented 49 percent of the total HECS payments required to be made to the sector. The balance of the payments required to be made were funded from a Commonwealth contribution of AU$870 million. The accumulated HECS debt on June 30, 2003 was estimated to be AU$9.1 billion (DEST 2004:3).

Funds for research and research training are allocated through a variety of performance-based funding programs administered by the Department of Education, Science and Training or the Australian Research Council's peer-reviewed competitive grants. Universities receive research funding from a number of other agencies and schemes, such as the National Health and Medical Research Council (NHMRC) and Co-operative Research Centres (CRCs). Research policy and funding is discussed in more detail in the following section of the chapter that is a revision and update of Wood and Meek (2002).

Overreviewed and Underfunded?

A number of peak bodies have argued the need for a stable and predictable policy environment to support universities and their research and development endeavors. However, the sector has been, and continues to be, the subject of a number of wide-ranging government inquiries and reviews—the outcomes of which have been variable. The range of these inquiries can be illustrated by some of the following reports:

- "Priority Matters" (Stocker, 1997).
- "Learning for Life: Final Report Review of Higher Education Financing and Policy" (West, 1998).
- "Review of the Greater Commercialization and Self-Funding in Co-operative Research Centres Programme" (Mercer and Stocker, 1998).

Apart from inquiries and reviews that have directly involved the higher education sector, other initiatives also have the potential to impact the way in which the sector operates. Examples include the following reports:

- "Going for Growth: Business Programmes for Investment, Innovation and Export" (Mortimer, 1997).
- "The Global Information Economy—The Way Ahead." (Goldsworthy, 1997)
- Investing for Growth 1997 (Government responses to Mortimer and Goldsworthy reports).
- A Platform for consultation 1999 (Ralph, 1999, "Review of the Australian Business Taxation: A Tax System Redesigned").

Substantial resources are involved in the sector participating in and responding to the terms of references of such government inquiries and reviews. However, where there has been little in the way of policy direction or funding commitment resulting from some of these reviews, their value to the sector must be questioned (Wood and Meek, 2002).

A particular theme of more recent government reviews and discussion and policy papers has been the role of universities in innovation. These include the following reports:

- "The Virtuous Cycle: Working Together for Health and Medical Research, 1999" (Wills, 1999).
- "New Knowledge, New Opportunities: A Discussion Paper on Higher Education Research and Research Training" (Kemp, 1999a).
- "Knowledge and Innovation: A Policy Statement on Research and Research Training," December 1999 (Kemp, 1999b).
- Innovation Summit January 2000/Australian Science Capability Review—"The Chance to Change," November 2000 (DISR, 2000).
- "Backing Australia's Ability"—Government Response to the Batterham Reports (Howard, 2001).
- "The Capacity of Public Universities to Meet Australia's Higher Education Needs" (Senate Review, 2001).

A brief overview of the key issues and recommendations of these reports and papers is provided below.

"The Virtuous Cycle", the Final Report of the Strategic Review of Health and Medical Research (Wills 1999)

This Report was released on May 12, 1999. It made major recommendations about the level and manner of funding available to universities, hospitals, and other research organizations for medical and medical biotechnology research. Among the issues identified were the following: (i) increasing the level of public investment; (ii) better management of research; (iii) greater involvement with industry; (iv) development of priority-driven research that contributes directly to population health and evidence-based health care; and (v) the education and training of health and medical researchers.

"New Knowledge, New Opportunities"

In June 1999, a discussion paper on research and research training, "New Knowledge, New Opportunities," was released, which provided the basis for extensive community debate about the policy and funding framework for university research and research training (Kemp, 1999a).

This discussion paper identified several deficiencies in the current framework which limit the capacity of institutions to respond to the challenges of the emerging knowledge economy: funding incentives that do not sufficiently encourage diversity and excellence; poor connections between university research and the national innovation system; too little concentration by institutions on areas of relative strength; inadequate preparation of research graduates for employment; and unacceptable wastage of resources associated with low completion rates and long completion times of research graduates. A particular concern was with research training and the funding of PhD and research masters students. The key reforms proposed by this discussion paper included:

1. An enhanced role for the Australian Research Council.
2. Research infrastructure as a component of research grants. The preparation by universities of research and research training management plans.
3. A new university block funding program, the Institutional Grants Scheme, to support research and research training and to encourage institutional diversity.
4. An Australian Postgraduate Research Student Scheme based on *portable* HECS exempt scholarships for research degree students.

"Knowledge and Innovation"

The government released its policy statement on research and research training "Knowledge and Innovation" in December 1999 (Kemp, 1999b). Major changes to the policy and funding framework for higher education research in Australia were identified in the policy statement. These included:

1. A strengthened Australian Research Council (ARC) and an invigorated national competitive grants system.
2. Performance-based funding for research student places and research activity in universities, with transitional arrangements for regional institutions.
3. The establishment of a broad quality verification framework supported by research and research training management plans.
4. A collaborative research programme to address the needs of rural and regional communities (Kemp, 1999b).

The most important recommendation of the White Paper for research management within universities concerns increased competition over research funding, particularly with respect to funding for PhD and research masters students.

"Knowledge and Innovation" instituted two new performance-based block funding schemes, intended to "reward those institutions that provide high quality research training environments and support excellent and diverse research activities." The Institutional Grants Scheme (IGS) supports the "general fabric of institutions' research and research training activities, and assist institutions in responding flexibly to their environment in accordance with their own strategic judgements" (Gallagher, 2000). The scheme absorbs the funding previously allocated for the Research Quantum and the Small Grants Scheme. Infrastructure funding through the Research Infrastructure Block Grants (RIBG) Scheme has been retained.

The IGS is distributed to universities through a performance-based formula comprising research income (60 percent), publications (10 percent), and number of higher degree research student places (30 percent). The RIBG Scheme supports project-related research infrastructure costs and is distributed across institutions according to their share of Australian Competitive Grants income. The government considers that institutions are likely to be more outwardly focused in their research when research income from all sources is equally weighted, unlike previous arrangements that gave lesser weight to income received from industry (Gallagher, 2000).

Funding for the Research Training Scheme (RTS) is also allocated through a performance-based formula. Institutions attract a number of scholarship places based on their performance through a formula comprising three elements: (i) numbers of all research students completing their degree (50 percent); (ii) research income (40 percent); and (iii) a publications measure (10 percent).

Innovation Summit/Australian Science Capability Review

Further development of a framework for higher education research has been assisted by the chief scientist's Review of the Science Base and the National Innovation Summit, announced by the then minister for Industry, Science and Resources, the Hon. Nick Minchin. The summit was held in early 2000 and was organized by the Business Council of Australia and the then Department of Industry, Science and Resources. The aim of the summit was to identify and develop a consensus on clear strategies for government, industry, and the research community to encourage future economic growth and improve Australia's competitiveness and innovation capacities.

The summit was supported by six Working Groups that focused on particular critical innovation issues. The Working Groups examined areas such as industrial innovation; intellectual property management; the human dimension of innovation; institutional structures and interfaces; innovation and incentives; and resource and infrastructure consolidation and cooperation (Wood and Meek, 2002).

Based on the Australian Science Capability Review the chief scientist presented a discussion paper in August 2000 entitled "The Chance to Change," Department of Industry, Science and Resources (DISR, 2000). The recommendations from this discussion paper and the resulting final report released in November 2000 centered around three themes: (i) culture; (ii) ideas; and (iii) commercialization. The principal recommendations included

- doubling the number of Australian Postdoctoral Fellows;
- providing 200 HECS scholarships for students undertaking science/education qualifications and 300 for students in maths/physics/chemistry;
- increased funding for the ARC and for university research infrastructure;
- testing a national site license concept between Higher Education Institutes (HEIs) and publishers to try and keep prices down;
- expansion of the CRC program; and
- more strategic approaches by universities and government-funded research agencies to the management of intellectual property.

To ensure that the recommendations of the review accorded with government and community objectives, an Implementation Committee was proposed (Wood and Meek, 2002).

"Backing Australia's Ability"

At the beginning of 2001, in a package entitled "Backing Australia's Ability" the federal government announced its AU$2.9 billion 5-year strategy to boost innovation. The strategy builds on a number of other government initiatives mentioned above. The main measures of the innovation plan can be summarized as follows:

- AU$995m HECS-style loan scheme for 240,000 postgraduate students (which Kemp has indicated could be capped).
- Twenty-five Federation Fellowships for top researchers, worth AU$225,000 a year for 5 years.
- New 175 percent tax concession for additional R&D: AU$460 million (all spending figures are total over 5 years).
- Existing 125 percent tax concession tightened to save AU$345 million.
- New 37.5 cents in the dollar R&D tax rebate for small companies: AU$13 million.
- Australian Research Council grants funding doubled: AU$736 million.
- Boost for research equipment, libraries, and laboratories: AU$583 million.
- R&D Start Programme continued for small and medium businesses: AU$535 million.
- Co-operative Research Centres Programme expanded: AU$227 million.

- Centers of Excellence in biotechnology and information technology: AU$176 million.
- Major national research facilities: AU$155 million.
- Twenty one thousand new full-time university places over 5 years in maths, science, and IT: AU$151 million.
- Foster science, maths, and technical skills in government schools: AU$130 million.

The government's additional investment was planned to achieve the following outcomes: "generating new ideas, developing ideas into products, and developing a highly skilled workforce. It strongly emphasized greater involvement of industry in research, to encourage more effective take-up of the results of research, and substantially increased commercial application" (AVCC, 2003b:3).

In addition to government commitment, The "Backing Australia's Ability" Plan also requires the states and business and research institutions to spend AU$6 billion over the same period to attract its grants and incentives.

Though the Innovation Strategy was welcomed by many in the public and private sectors, there was the question of whether the financial commitment would be sufficient to offset the substantial funding cutbacks made to the higher education sector since 1996. Despite this being "the largest commitment to innovation ever made by an Australian Government" (Howard, 2001), it only spent AU$159.4 million in its first year of 2001–2002. Much of the funding did not begin to flow for 2 or 3 years after the announcement—with AU$946.6 million to be outlaid in 2005–2006.

The Capacity of Public Universities to Meet Australia's Higher Education Needs—Senate Review 2001

The Senate review of higher education was announced at the end of 2000. The terms of reference were extremely broad and included the following: (i) the adequacy of current funding arrangements with respect to the capacity of universities to manage and serve increasing demand, institutional autonomy and flexibility, and the quality and diversity of teaching and research; (ii) the effect of increasing reliance on private funding and market behavior on the sector's ability to meet Australia's education, training and research needs, and the quality and diversity of education; (iii) the capacity of public universities to contribute to economic growth; and (iv) the regulation of the higher education sector in the global environment.

The review received more than 300 submissions and collected evidence at a number of public hearings. Recommendation 1 of the report states that "the government end the funding crisis in higher education by adopting designated Commonwealth programmes involving significant expansion in public investment in the higher education system over a ten year period." However, the receptiveness of the government to arguments that the sector needs additional funding has been minimal (Wood and Meek, 2002).

"Higher Education at the Crossroads"

Throughout 2002 the federal government conducted a review of Australian universities under the banner "Higher Education at the Crossroads." Despite a number of position papers and numerous submissions from the sector, government policy was merely announced as a fait accompli as part of the 2003 budget statement.

The package of higher education reforms was entitled: "Universities: Backing Australia's Future." Though there is commitment of some new money, basically the policy continues the trend toward greater privatization of higher education funding through increasing tuition fees, allowing institutions to set their own fees (within a range), and allowing institutions to enroll a greater number of full fee-paying domestic undergraduate students. After protracted debate and a number of amendments to the recommendations, the following recommendations were accepted by the Australian parliament in December 2003:

- More than 34,000 new Commonwealth-supported places.
- Increasing the Commonwealth contribution per student place by 2.5 per cent from 2005, building to a 7.5 per cent increase by 2007, conditional on institutions providing staff with genuine choice of industrial agreements and adherence to the National Governance Protocols which are designed to encourage efficiency, productivity and accountability in the sector.
- Providing greater support for regional campuses.
- Raising the repayment threshold under the Higher Education Contribution Scheme-Higher Education Loan Programme (HECS-HELP) from AU$.24,365 in 2002–2003 to AU$.35,000 in 2004–2005 (AU$.36,184 in 2005–2006) which will significantly improve the financial position of many graduates with lower incomes.
- AU$.327 million for two new scholarship programmes over the next five years to assist students with education and accommodation costs.
- More than AU$.50 million in additional funds over five years to support a range of equity initiatives.
- From 2005, universities will be able to set student fees within a range from 0 to a maximum 25 per cent above the current Higher Education Contribution Scheme (HECS) rates.
- Increasing the maximum number of Australian fee-paying students (with the exception of medicine) from 25 to 35 per cent of a total course cohort.
- A new programme to enable all full fee paying students undertaking an award programme at an eligible institution to borrow the amount of their tuition fees from the Commonwealth. These loans will be subject to the same repayment arrangements as under the HECS-HELP programme.
- Providing student learning entitlements to cover the duration of a Commonwealth-supported student's course for up to seven years with flexibility for an extension in the case of longer courses.
- Providing places for the national priority areas of nursing and teaching and special fee arrangements to encourage people to enroll in these fields.
- A new learning and teaching performance fund will be introduced from 2006 to reward institutions that best demonstrate excellence in learning and teaching. A total of AU$.251 million will be allocated under the fund between 2006 and 2008.
- A new national institute for learning and teaching in higher education will be established with ongoing annual funding of AU$.22 million from 2006.

- A total of AU$.83 million will be allocated between 2006 and 2008 under the new workplace productivity programme to encourage improvements in workplace productivity.
- Additional funding of AU$.4 million over five years on quality initiatives including additional funding to enhance the operations of the Australian Universities Quality Agency in relation to offshore audits.
- A new collaboration and structural reform fund will be established for three years from 2005 to encourage innovation and collaboration within the sector.
- Approximately AU$.40 million in transitional funding to ensure that no institution is disadvantaged under the new funding arrangements.

According to the minister, the recommendations will result in an increase in public investment in the sector of AU$2.6 billion over the next 5 years and AU$11 billion over the next 10 years (DEST, 2004:3). But as with the funding commitments in "Backing Australia's Ability," most of the funding increases come at the end rather than the beginning of the periods identified.

The "Higher Education at the Crossroads" review and recommendations had little to do with research per se. The minister of Education, Science and Training in recognizing this announced in 2003 six additional reviews of research: (i) mapping Australia's science and innovation; (ii) research collaboration between universities and publicly funded agencies; (iii) evaluation of knowledge and innovation; (iv) national strategy on research infrastructure; (v) review of Backing Australia's Ability; and (vi) evaluation of cooperative research centers. The reports of several of these reviews were released in late 2003 or early 2004. But the recommendations appear to do little to disturb the status quo, and at the time of writing the government had yet to declare its position on the recommendations. Some of the outcomes of the report "Mapping Australia's Science and Innovation" are outlined in the following section of this chapter.

It is interesting to note that the "Evaluation of Knowledge and Innovation" (Fell, 2004) recommends "[t]hat the government provide increased funding to allow universities to carry out their responsibilities to renew and enhance their institutional research infrastructure, to develop their own strategic research focus and to properly carry out competitively awarded research projects." Other recommendations included consideration of research quality assessment, possibly using a modified form of the United Kingdom's Research Assessment Exercise (RAE).

Inquiry into Higher Education Funding and Regulatory Legislation—Senate Review 2003

In June 2003 the Senate referred to its Employment, Workplace Relations and Education References Committee the task of inquiring into the impact of the "Universities: Backing Australia's Future" recommendations. The committee established a subcommittee, which held numerous public meetings and received more than 480 submissions before reporting to Parliament in November 2003. The Senate report was entitled "Hacking Australia's Future, Threats to Institutional Autonomy, Academic Freedom and Student Choice in Australian Higher Education." The report was highly critical of many of the governments' proposed policies, recommending that "[t]he bill is so badly flawed, at both a philosophical and technical level that it should not be given a second reading." But as discussed above, the government was able to muster the number to gets its legislation passed following more or less minor amendments.

National Research Priorities

At the beginning of 2002, the government announced, as a result of a "consultation" process that was far from transparent, that a portion (33 percent) of the Australian Research Council's (the largest nonmedical research funding agency in Australia) funding would be targeted to research in the following four priority areas: nanomaterials and biomaterials, genome/phenome research, complex/intelligent systems, and photon science and technology.

In May 2002, the government instituted a review process to further set national research priorities for government-funded research in the areas of science and engineering. According to government, the priorities "will highlight research areas of particular importance to Australia's economy and society, where a whole-of-Government focus has the potential to improve research, and broaden policy outcomes" (DEST, 2002:1). The priorities announced at the end of 2002 are

- an environmentally sustainable Australia;
- promoting and maintaining good health;
- frontier technologies for building and transforming Australian industries; and
- safeguarding Australia.

These priorities subsume the ARC research priorities mentioned above. When the priority review process was first initiated, the intention was to follow the research priority-setting exercise in the sciences and engineering with one in the social sciences and humanities. But that did not happen. Rather, subgoals for each priority area were written in such a way that the social sciences and humanities could be incorporated. Nonetheless, while broad in scope, the priorities are "hard-science"-oriented and mainly emphasize areas of immediate economic relevance. The research priorities are applicable across all Commonwealth's research agencies and funding bodies.

Analysis of Outcomes of Policy Implementations and Issues Raised

It is quite difficult for several reasons to assess the impact and outcomes of policy on higher education and its research effort. Data is often sporadic, out of date, and difficult to obtain. But more importantly, the effects of particular policies often take a considerable amount of time to appear.

This problem is exacerbated when the implementation of the funding dimensions of particular polices is relegated to a relatively distant future—which is the case for the set of policies contained in both "Backing Australia's Abilities" and "Our Universities: Backing Australia's Future." The stark effects of the government's 1996 financial cutbacks to higher education did not become blatantly apparent until the end of that decade. The impact of new money committed through "Backing

Australia's Abilities" is only starting to emerge. But while the analysis of higher education and research policies are necessarily complex, they are also vital given the importance of the sector to the nation's economic and social well-being. Given the caveats mentioned above, this section of the chapter will attempt to summarize what appear to be some of the key trends and issues in Australian research policy and effort.

The report "Mapping Australian Science and Innovation" lists a number of weaknesses of Australian Science (DEST, 2003a). These include the following:

- Australia's scientific standing in the world may be at risk and, in general, Australian science and patented technology has limited visibility and impact on the development of world technologies.
- Business innovation involving R&D and development of new technology remains low by international standards.
- Investment in the development of strategic Information and Communications Technology (ICT) capability is low, which may weaken the innovation base and the future competitiveness of the economy.
- Australia's commercialisation record . . . remains low compared to other countries and is uneven within and across different research sectors. Continuing barriers to commercialisation include lack of access to early stage capital, a shortage of management and entrepreneurial skills and lack of fully effective links between researchers and industry.
- Challenges remain in fostering science and innovation collaboration and linkages, especially between publicly funded research providers and industry.
- Australia's research infrastructure is under pressure in terms of investment and maintenance, and in leveraging access to international research infrastructure in an environment of increasing scale, costs and technical complexity.
- The long-term sustainability of Australia's skills base in the enabling sciences is under pressure in some areas with declines in participation in most science subjects in Year 12 and in S&T subjects at the undergraduate level at university.
- Availability of innovation skills and cultural attitudes towards innovation limit Australia's innovation potential.
- While total gross expenditure on R&D as a proportion of GDP is now some 50 per cent higher than in 1981, Australia continues to rank towards the bottom of OECD countries in terms of R&D investment.
- Government support for business R&D is low by international standards, being less than half that of the leading OECD countries.

Besides being overreviewed, the single most fundamental issue facing Australian higher education in general and research specifically remains, not surprisingly, funding. However, money is not the only issue. At the heart of the problem is the question of whether Australia is to have a publicly supported, publicly subsidized, or fully private higher education system. Some have argued that little will be achieved with respect to funding until government agrees to restore full supplementation of operating grants for increases in academic salary. While the government in the present round of reforms has committed some new money to the sector, most of it will be absorbed by the present round of salary increases as a result of enterprise bargaining. Moreover, as indicated above, the government's main funding reform has been to shift even more of the burden to the student consumer. But student fees will not support an increased research effort. In fact, with an ever worsening staff/student ratio, in some universities money earned through research effort is actually subsidizing teaching through payment of staff salaries.

Australian higher education faces fundamental structural and long-term funding issues. The longer research infrastructure is allowed to decline, the more difficult it becomes for the nation to recover its R&D standing relative to the rest of the world. A past president of the Australian Vice-Chancellors' Committee observed that "the pace of change in public investment in universities is such that if our universities get too far behind those in other countries we will not catch up" and also raised the concern that "Australia will become an importer of knowledge and an exporter of talent and we will have too few educated personnel locally to add value to the efforts of others let alone enough to produce from our own." (Chubb, 2000:3). For a number of historical and structural reasons, the Australian research effort is more dependent on public support than most OECD nations. However, the ideological commitment of the government has been to the market and privatization.

A deep issue in Australian higher education research is the connection between teaching and research. On the one hand no country can afford to fund all of its universities as if they were world-class research-intensive institutions. On the other hand, there are those who argue that all university teaching must be informed by research. Moreover, each institution has its own special arguments about why it should be recognized as a leading research university (whether or not the facts support such arguments). The collapse of the binary system of higher education in the early 1990s has exacerbated this problem. The introduction of new research performance-based funding measures (RTS and IGS) mentioned above are designed to concentrate research funding on the research performers. It is too early to tell whether the policies will have the desired effect since up to 2005 a cap has been placed on how much funding individual universities can loose or gain through the application of the policies. But in the longer term, more radical policies may be necessary.

Appropriately, government has instituted a number of policies to boost business investment in R&D. "Mapping Australian Science and Innovation" (DEST 2003b:367) pointed out that "Australia is the only country in which business funding of research and development as a percentage of GDP is lower than Government funding of research and development as a percentage of GDP." The review in a background paper also observed that a key OECD finding is that "rapid growth in research and development is largely driven by increases in business-performed research and development" (AVCC, 2003b:10). Given the country's history of investment in R&D, it is probably necessary to attempt to increase the share coming from business and industry. But this should complement, not diminish, the investment from other sectors, particularly government.

With research policy strongly based on principles of concentration and selectivity, it is hardly surprising that the government would wish to set national research priority areas. The danger here, however, is that if funding becomes progressively concentrated in priority areas, innovation may be "straight jacketed." This is one of the dilemmas a small country with a limited science base faces. While the nation cannot adequately fund all kinds and aspects of modern research, it must maintain a broad enough science base to participate in advances in knowledge globally. According to the AVCC (2003b:22), "The key issue is plurality: as a nation we need to support a range of research, and do so by a number of different means. Allowing any single approach to dominate would inevitably result in a diminished overall research capacity and a weaker national innovation system. The impact of research prioritization should be restructured to recognize this fact."

Another aspect of priority setting is the prominence given to science and engineering at the expense of the social sciences and humanities. The present round of priorities gives little more than lip service to the social sciences. Much of the present thinking is based on the assumption that worthwhile research means commercialization and commercialization means science and technology. Again, a more balanced approach is necessary. The social sciences have much to add, particularly to the nation's social and cultural prosperity. They also have an important role to play as critic of the environmental and social consequences of scientific and technologically driven development. But with an increasing emphasis on commercialization, the role of the university of "speaking truth to power" may be lost sight of. There is some evidence to suggest that this is a significant problem in the USA higher education sector Newman et al. (2004). Even the AVCC (2003b:12) agrees that "[r]ecent priority setting in research has underrated the contribution made by the social science and humanities."

The AVCC states that

[t]he research base must include the humanities and social sciences. The focus in recent years on Science, Engineering and Technology (SET) has marginalized areas of research and scholarship which play an important role in society, and which provide the vital critical and creative underpinnings of many other disciplines. The social sciences and humanities provide an understanding of ourselves and of the human and natural world around us, and work with the sciences towards resolving the full spectrum of problems and challenges which confront us. (AVCC, 2003b:12)

Related to the issue of priority setting are the emphases placed on pure basic research relative to applied and developmental research. Both government and institutional management alike have been very interested in the commercialization of research outcomes. This has resulted in a shift of funding over the years from pure basic to applied research, as is depicted in table 4.8. The linear view of scientific innovation no longer has credibility.

Nonetheless, if basic, "blue-sky" research is progressively diminished, the fountain of ideas and advances in knowledge that feeds other forms of research and technological innovation may dry up as well (AVCC, 2003b:7).

The emphasis on applied research reflects the concern by both government and institutional leaders that research outcomes are commercialized, which in turn leads to the funding of the type of research most likely to achieve this result. This appears to have resulted in a sharp decline in pure basic research, as is reflected in table 4.9.

Society remains the major research category with respect to socioeconomic objective of research, partially due to the fact that health research is classified under society. The category of "economic development" is steadily increasing, while the most

Table 4.8 University expenditure on research and experimental development by type of research activity, 1988–2000 (annual expenditures are in respective year prices)

Type of research activity	1988 (%)	1990 (%)	1992 (%)	1996 (%)	1998 (%)	2000 (%)
Pure basic research	38.0	41.0	40.0	34.1	33.5	30.5
Strategic basic research	24.0	22.0	24.0	25.0	25.4	24.0
Applied research	31.0	31.0	30.0	34.7	35.0	37.8
Experimental development	7.0	6.0	6.0	6.2	6,1	7.7
Total (AU$ million)	1,076.8	1,350.8	1,695.2	2,307.6	2,600.2	2,774.6

Source: AVCC, 2003a:7.

Table 4.9 Socioeconomic objective of research by type of funds (percentages)

Percentage of total HERD for SEO	Economic development			Society			Environment			Nonoriented research		
	1996 (%)	1998 (%)	2000 (%)	1996 (%)	1998 (%)	2000 (%)	1996 (%)	1998 (%)	2000 (%)	1996 (%)	1998 (%)	2000 (%)
All sources	21	23	29	25	27	40	7	7	6	46	42	25
Commonwealth National Competitive Grants	21	24	27	25	26	39	8	7	6	46	43	28
State and local Government	21	27	31	51	47	48	10	9	12	18	17	10
Business	43	42	44	21	22	32	9	10	7	26	25	16
General university funds	18	22	28	24	27	40	7	7	5	50	44	27
Overseas	23	26	27	36	32	47	6	4	5	33	38	21

Source: AVCC, 2002:3.

alarming trend is "the sharp decline in '*Non-oriented Research*,' or what used to be classified as '*Advancement of Knowledge*.' Fields that fall into the nonoriented research category include (i) mathematical sciences; (ii) physical sciences; (iii) chemical sciences; (iv) earth sciences; (v) biological sciences; (vi) political science and public policy; (vii) studies in human society; (viii) and behavioral and cognitive sciences (AVCC, 2002:2).

Noting the decline in basic research, the AVCC (2003b:19) warns that "without a strong footing in pure basic research the national innovation system will run out of ideas—or have to import them, at increasing expense, from elsewhere. Secure and substantial investment in basic research is decidedly in the national interest."

Conclusion

To conclude, the argument in this chapter is that *there is not* one best approach to coordinating and funding university research at the national level. A number of competing demands must be balanced—*balance and plurality* are the key words. Moreover, while the public good nature of research must be recognized and supported, the fact remains that someone must pay for it. As research becomes more elaborate and expensive, policies of concentration and selectivity are necessary. With respect to research, government and universities alike must make choices. But the choices must be informed ones—not driven primarily by ideology—and take place within a set of parameters that will sustain the research endeavor in the long term. In this respect, the AVCC's (2003b:4) recent advice to government on a follow-up package to "Backing Australia's Ability" provides some useful guidelines. The key critical factors for research excellence and innovation listed by the AVCC are as follows:

- A pluralist research funding system that supports a dynamic range of research, with no single body or approach dominating to the detriment of innovation.
- Substantial and increasing investment in universities core research building capacity, based on the quality of each university's performance that enables each university to pursue its strategic objectives and support national needs.
- Major ongoing investment in research infrastructure.
- Effective incentives for business to commit effectively to research and innovation.
- A commitment to basic research as essential to innovation.
- A commitment to research that includes all knowledge areas to include the social sciences and humanities (as well as science, engineering and technology).

Australia has no choice but to further develop its national research capacity, a task in which the country's universities have a leading role to play. Government's task is to provide the policy context, including funding, to ensure that the universities continue to provide the innovation base necessary for the nation's economic and social advancement. Higher education contributes to the knowledge factory, but the products of that factory need to be seen in social and cultural as well as economic terms.

References and Works Consulted

AVCC (Australian Vice-Chancellors' Committee) (2000) "Our Universities: Our Future." Support Papers, Canberra, September.
—— (2002) "Major Points from the ABS Release: Research and Experimental Development Higher Education Organisations, Australia." Canberra, April.
—— (2003a) "Key Statistics on Higher Education: Research." Canberra.
—— (2003b) "Advancing Australia's Abilities: Foundations for the Future of Research in Australia." Canberra.
Barnett, R. (2000) *Realizing the University in an Age of Super-Complexity*. Buckingham: SRHE and OUP.
Chubb, I. (2000) "Our Universities: Our Future." An AVCC discussion paper, AVCC, Canberra, December.
—— (2001) "When to be Average is to Fail." National Press Club Address, Sydney, March 14, 2001.
Clark, B. (1998) *Creating Entrepreneurial Universities: Organisational Pathways of Transformation*. London: Pergamon Press.

Dawkins, J. D. (1988) "Higher Education: A Policy Statement." White Paper, AGPS, Canberra.
DISR (Department of Industry, Science and Resources) (2000) "The Chance to Change." Discussion paper by the Chief Scientist, Batterham, Canberra, November.
—— (2002) "The Framework for Setting National Research Priorities." Canberra.
DEST (Department of Education, Science and Training) (2003a) "Mapping Australian Science and Innovation." Summary Report. Canberra: McMillan Printing Group.
—— (2003b) "Mapping Australian Science and Innovation." Final Report. Canberra: McMillan Printing Group.
—— (2003c) "Australian Science and Innovation System: A Statistical Snapshot." Canberra.
—— (2004) "Higher Education Report for the 2004–2006 Triennium." Canberra.
Fell, C. (2004) "Evaluation of Knowledge and Innovation Reforms Consultation Report." Canberra: Commonwealth of Australia.
Gallagher, M. (2000) *The Emergence of Entrepreneurial Public Universities in Australia*. Higher Education Division Occasional Paper Series 00/E. Canberra: DETYA.
Goldsworthy, A. (1997) "The Global Information Economy—The Way Ahead." IIFC Report of the Information Industry Taskforce, Australian Department of Industry, Science and Resources.
Hayden, M. (2003) "An Australian Perspective on System-Level Strategic Performance Indicators for Higher Education." In A. Yonezawa and F. Kaiser, eds., *System-Level and Strategic Indicators for Monitoring Higher Education in the 21st Century*. Bucharest: UNESCO, chapter 25.
Howard, J. (2001) "Backing Australia's Ability." Commonwealth of Australia, January, 2001.
Jackson, J. (2002) "The Marketing of University Courses under Sections 52 and 53 of the Trade Practices Act 1974 (Cth)." Institute for Scientific Information (2001). *Southern Cross University Law Review*, 6:106–132, http://sunweb.isinet.com/isi/index.html (accessed July 31, 2005).
Kemp, D. (1999a) "New Knowledge, New Opportunities: A Discussion Paper on Higher Education Research and Research Training." Canberra, AusInfo, June.
—— (1999b) "Knowledge and Innovation. A Policy Statement on Research and Research Training." Canberra, AusInfo, December.
Meek, V. L. (2002) "Evaluation of Higher Education Programmes and Initiatives in Other States and Territories which may Inform the Development of Policy in Victoria." Report. Office of Higher Education, Victoria Department of Education and Training, 1–153.
Mercer, D., and J. Stocker (1998) "Review of Greater Commercialization and Self-Funding in the Co-operative Research Centres Programme." Department of Industry, Science and Tourism, Canberra.
Mortimer, D. (1997) *Going for Growth Business Programmes for Investment, Innovation and Export*. Canberra: AGPS.
Neave, G. (2002) "The Stakeholder Society." In J. File and L. Goedegebuure, eds., *Enschede: Center for Higher Education Policy Studies (CHEPS) Inaugurals*. The Netherlands: Centre for Higher Education Policy Studies, University of Twente, Enschede.
Newman, F., L. Couturier, and J. Scarry (2004) *The future of Higher Education: m Rhetoric, Reality and the Risk of the Market*. San Francisco: Jessey-Bass.
Phillips, D. et al. (2002) "Independent Study of the Higher Education Review." Sage 1 Report, Byron Bay, NSW, Phillips Curran, http://www.kpac.biz (accessed July 31, 2005).
Ramsey, A. (2001) "Spaghetti and Meatballs Land Kim in Soup." *The Sydney Morning Herald*, Sydney, July 4.
Ralph, J. (1999) "Review of Australian Business Taxation: A Tax System Redesigned." (Ralph—chair), Treasury, Canberra, July 1999.
Scott, P. (1997) "The Changing Role of the University in the Production of New Knowledge." *Tertiary Education and Management*, 3(1).

Scott, P. ed. (1998) *The Globalization of Higher Education*. Buckingham: SRHE and the Open University Press.

—— (2003) "Changing Players in a Knowledge Society." In G. Breton and M. Lambert, eds., *Universities and Globalization*. Paris: UNESCO, 211–222.

Senate Review (2001) "The Capacity of Public Universities to Meet Australia's Higher Education Needs." Report. Canberra: Commonwealth of Australia.

—— (2003) "Employment, Workplace Relations, Small Business and Education References Committee: Hacking Australia's Future, Threats to Institutional Autonomy, Academic Freedom and Student Choice in Australian Higher Education." Commonwealth of Australia, Canberra.

Stocker, J. (1997) "Priority Matters." A report to the Minister for Science and Technology, on Arrangements for Commonwealth Science and Technology, by the Chief Scientist, Professor John Stocker, Canberra, AGPS, June.

The Economist (1997) "Inside the Knowledge Factory." October 2, 1997, http://www.economist.com (accessed July 31, 2005).

The Sydney Morning Herald (2004) Canberra, April 22, 2004.

West, R. (1998) "Learning for Life: Final Report of the Review of Higher Education, Financing and Policy." Report. Canberra: Department of Employment, Education, Training and Youth Affairs.

Wills, P. (1999) "The Virtuous Cycle: Working Together for Health and Medical Research." Report. *Health and Medical Research Strategic Review* (Wills—chair). Canberra: Commonwealth of Australia.

—— (2001) "The National Investment in Science, Research and Education." Address to the National Press Club, Canberra, August 21, 2001.

Wood, F. Q., and V. L. Meek (2002) "Over-Reviewed and Under-Funded? The Evolving Policy Context of Australian Higher Education Research and Development." *Journal of Higher Education Policy and Management*, 24(1):7–25.

Chapter Five
Policy Debate on Research in Universities in China

Wei Yu

Introduction

China is advancing and changing rapidly in many areas, one of which is the development of the higher education system. There is ongoing debate on how universities could advantageously conduct their research in conjunction with their teaching activities, and especially on laying the foundations of high-tech companies and science parks related to universities. The lessons learned from experience—and the successes thereof—are reflected in China's endeavors to seek suitable methods of further development.

An Overview of the Reforms and Development in Higher Education in China Since the 1990s

As from the beginning of the 1990s the development of the higher education system in China witnessed two stages, from the early 1990s to 1998 and thereafter from 1998 to the beginning of the twenty-first century. Though linked to each other in several ways, each stage has its own characteristics.

From the early 1990s to 1998, the Ministry of Education mapped out the line of the development of higher education in China according to "The Outline of Educational Reforms and Development in China" (MOE, 1993). From the late 1980s, the ministry assigned a special group to draw up this outline based on its own investigations, which was officially approved in February 1993 by the Central Committee of the CPC and the State Council. This stage was characterized by the ministry's strict control over the scale of the development in various universities, including the size of enrollment for both undergraduates and postgraduates, with the explicit stipulation that universities should place emphasis on overall reforms and the *enhancement of quality*. The reforms and development of this stage may be summarized as follows:

1. The aim was to have strict control over the total quantity of college enrollment. The aforementioned "Outline" stipulated that the total number of students of

universities and other higher education institutes (HEIs) should be about 6,300,000 and the entrance rate for those around the 18–21 age group was expected to be 8 percent; later the total number of students was revised to approximately 7,300,000. In 1998, the actual total enrollment of college students was 6,230,900.
2. From the early 1990s, a series of reforms in HEI were carried out, including administrative reforms, teaching reforms, tuition reforms, and the promotion of private and Sino-foreign joint institutions of higher learning. These have been full-scale and in-depth reforms with far-reaching effects. For instance, colleges began to charge tuition fees in 1993 for the first time since the founding of the People's Republic of China (PRC). Initially this happened at two universities, South-Eastern University and Shanghai Foreign Language College, then smoothly extended to all Chinese higher education institutes (HEIs). Another example is the promulgation of the "Interim Regulations" of the People's Republic of China on Chinese-Foreign Co-operation in Running Schools in 1995. The higher education institutes founded by nongovernmental organizations (NGOs) and private sectors were also established at this period.
3. The implementation of "Project 211" took place in 1995. It was the Chinese government's new endeavor that aimed at strengthening about 100 universities and key disciplinary areas, put forward as a national priority for the twenty-first century. Included in the project were 98 institutions of higher learning. Universities with a relatively higher level of research capacity were and are still all supported by this project. Universities in western China are also taken into consideration. The funding required for "Project 211" is generated through a cofinancing mechanism involving the state, the local governments, and higher education institutions concerned. Investment in the project will total 10,894 billion RMB, of which 2,755 billion will come from the central government, and the rest will be raised by the local governments and the universities involved. Project 211 funds a total of 602 research programs of key disciplinary areas plus some public service items like digital libraries connecting key universities, Chinese Education and Research Networks. The implementation of Project 211 plays an important role in upgrading the academic level of the universities concerned.
4. Project 211 also provides an opportunity to promote cooperation between universities and local governments concerned in their joint effort to reform the administrative setup of institutions of higher learning. It is for this reason that the Ministry of Education (MOE) has put forward a policy calling on the parties concerned to pool their efforts and build up the activities of universities at the academic level.

Reforms carried out at this stage were stable, with both the teaching quality and academic atmosphere exhibiting marked improvement. The major problem was that the state funding for higher education was far from sufficient, and MOE was criticized for its "overstrict control" over the scale of higher education development, and its "overcentralized" administration over universities. Some even branded the ministry as "the last stronghold of planned economy in China."

Since 1998 China's higher education system has been taking considerable strides. Madam Chen Zhili, the new education minister, presided over the drawing up of "The Action Plan of the Revitalization of China's Education to Meet the Needs of the 21st Century" (1999), which was approved by the State Council in January 1999. And the Ministry of Education began to focus its attention on the implementation of the "Scheme for Trans-Century Education Reforms and Development" as detailed in the "Action Plan." Development of higher education at this stage is characterized by what follows:

1. A drastic increase in the total quantity of college enrollment. In 1999 the Science and Education Leading Group of the central government decided to increase the size of college enrollment (National Centre for Education Development Research, 2002). As a result of the continuous increase from 1999 to 2004, the total number of college students in 2004 exceeded 20 million, making the system the largest in the world. The gross enrollment ratio (GER) had exceeded 19 percent (see figure 5.1).
2. At this stage, university administration reforms have been basically accomplished, and major changes have been made in the former system of practices during the period of planned economy, in which universities were under the administration of related ministries. In place of the old system, a two-level administrative system has been established, which involves the participation of both the central government and provincial governments concerned, with the administration at the provincial level taking major responsibilities. More than 900 universities have been involved in the cooperation, unification, or merger

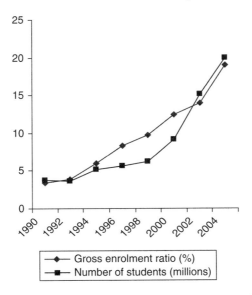

Figure 5.1 Change in the gross enrollment ratio (GER) and the total number of students (1990–2004)

Source: MOE, 2004.

programs of various forms. As a result, a number of "jumbo-sized" comprehensive universities have been established. For instance, Jilin University, located in Changchun City, Jilin Province, is currently the largest university in China. It is a combination of 7 colleges, with a current enrollment of 60,000 students, and approximately 20,000 faculty members.

3. More money is to be invested in higher education. On the hundredth anniversary of Peking University in May 1998, the then president Jiang Zemin announced that China would build the best first-class universities in the world (Jiang Zhe Ming, 1998). After that, the Ministry of Education embarked on the "985 Project," named after the date of President Jiang's announcement. Accordingly, nine "first-echelon" universities, including Peking University and Tsinghua University, obtained intensive support, followed by more than 20 "second-echelon" ones aimed at becoming "higher-level" universities of the world. Of the 14,469 billion RMB appropriated by central finance in fiscal years 1999–2002 and used for carrying out The Action Plan for Vitalizing Education of the twenty-first century, 12.612 billion RMB was allocated to those universities mentioned above (MOE, 2002).

4. Social services of higher education, like scientific research and technology transfer, are active. The money put into scientific research in universities in 2004 was five times more than that of 1996 (Zhou Ji, 2004).

5. Along with the rapid expansion of the scale of higher education, some new challenges appear:

   i. The development of vocational education and elementary education in rural areas is relatively slow with respect to the requirement of economic and social development in China.
   ii. Some administrative problems result from the expansion and reorganization of the institutes of higher learning. Coupled with inadequacy of teaching staff and reduction of per capita money for students, the quality of higher education has been a great concern for the nation as a whole.
   iii. College graduates' employment has become a "hot topic." In recent years the vice-premier would preside over a special meeting to discuss this topic. In 2004 the official data for college gradates' employment was about 75 percent.
   iv. Some universities, irrespective of their present conditions, compete for aiming at becoming first-class or prestigious universities of the world so much so that they blindly invest in building ambitious campuses and set up some new disciplines that duplicate existing programs.

All circles of society have mixed judgments on the rapid expansion and merger of universities after 1999. The pressures and impetus for expanding higher education come mainly from domestic political, economic, social, and cultural considerations, sometimes also from international influences, especially those aid-providing countries and influential international organizations, like the World Bank (WB) and the Organisation for Economic Co-operation and Development (OECD). The factors mentioned above partly contribute to the wave of rapid expansion of higher education.

However, this chapter holds that the political willingness and instinctive decisions of those in charge are the decisive factors. It takes time to assess such a far-reaching decision, just as an old Chinese saying goes, "It takes ten years to grow trees but a hundred years to rear people."

This overview serves as background information for further discussion of the policy concerning the development of science and technology in Chinese institutions of higher learning.

Some Definitions

As far as universities are concerned, their primary goal is to cultivate talents, which entails not only *teaching* but also many other activities including *research* work. Apart from the cultivation of talents, universities should also contribute to scientific progress and provide the society with some direct services. All this cannot be separated from research work. Research mentioned here includes research and development (R&D). According to the definition given by UNESCO, "science includes natural science and social science." Restricted by its length and the author's knowledge, this chapter mainly deals with the R&D of natural science. According to the definitions given by the Ministry of Science and Technology (MOST) of China, some terms on R&D involved in this chapter are defined as follows (MOST, 2001):

Research and Development (R&D): it refers to the systematic and creative activities (including basic research, application research and development) in the field of science and technology, aiming at increasing the total amount of knowledge and applying the knowledge to creating new applications.

Basic Research: it refers to the experimental or theoretical research aimed to acquire new knowledge about the basic principles, like those revealing basic laws of things, governing phenomena and observable facts, but not aimed at any specific or particular application.

Application Research: it also refers to creative research for acquiring new knowledge, but it aims at a specific target, that is, to explore the possible applications of basic research or the new methods or approaches that may be used to reach a certain goal.

Development: it refers to the systematic work of applying the knowledge acquired from basic research, application research and actual experience, so that (i) new products, materials and devices can be developed, (ii) new techniques, systems and services can be established, and all the above having been developed or established can be substantially improved.

A Summary of the Scientific Research in Chinese Institutions of Higher Learning

Scientific research in Chinese institutions of higher learning falls into three stages: (i) initial and trying stage from 1947 to 1976; (ii) recovery and development stage from 1977 to 1984; and (iii) reform and development stage from 1985 onward.

In the early 1950s, the Chinese government, following the former Soviet Union model, separated scientific research from universities and set up a separate Chinese Academy of Science. This was why scientific research in universities was weak at that time. After years of effort, in the early 1960s, the scientific research in Chinese institutions of higher learning began to be regulated by the government, which appropriated money on a regular basis. Unfortunately, the Cultural Revolution beginning from 1966 proved to be a disaster to Chinese universities, setting back its scientific research activities severely.

The year 1977 was a transitional year for Chinese institutions of higher learning, when the late Mr. Deng Xiaoping pointed out that universities, especially the key ones, should play an important role in scientific research. After that, universities began to resume their scientific research. In 1981, China began to train its own masters and doctors.

In 1985, the Chinese government started reforms in its systems of science and education. Competition was introduced into the funding mechanism of scientific research, and about 100 national research centers supported financially by the central government were established in universities, which have quickly become the main forces behind nationwide scientific research, especially in the area of basic research. According to the statistics of 2002 (MOE, 2002; MOE, 1999; MOE, 2003a), of all the teaching staff in universities, those who were wholly devoted to scientific research totaled 286,000 (excluding hundreds of thousands of postgraduates), among whom there were 450 academicians, accounting for 33 percent of the total in the whole country. Over the past 2 decades, financial contributions to scientific research in universities has increased considerably, from 590 million in 1985 to 21.9 billion in 2002, an increase of more than 30 times (see figure 5.2). Of the research projects supported by the National Foundation of Natural Science, 70 percent are taken up by universities, and of national key research projects, 60 percent by universities, which also take up 30 percent of the national high-tech research programs. Of the three National Science Rewards, universities win half of the National Science Reward, one-third of the National Invention Reward, and one-fourth of the National High-Tech Progress Reward. Furthermore, they account for over 60 percent of the total research papers published at home and abroad. In terms of social science, universities are a major force. Take the year 2000 as an example (MOE, 2002).

Of the 23,446 research items of social science, universities were responsible for 20,301, about 86.5 percent of the total, with a financial investment amounting to 340 million. All the above demonstrates that universities gradually became a major force in national scientific research, particularly in basic research.

According to the data of "Bulletin of Statistics on Chinese Education in 2003" issued by the Ministry of Education (MOE, 2004), there are altogether 2,210 ordinary universities and institutions of higher learning for adults in China in 2003, of which 1,552 are ordinary universities. Among them, 644 can confer a bachelor's degree, 407 a master's degree, and 260 a doctor's degree. The universities that can train postgraduates, especially doctoral candidates, have to be more active in their research work.

Most of the research work of natural science in universities of China is concentrated in about 60 institutions, which have gathered approximately 70 percent of graduates and 90 percent of doctoral candidates as well as over 85 percent of doctoral

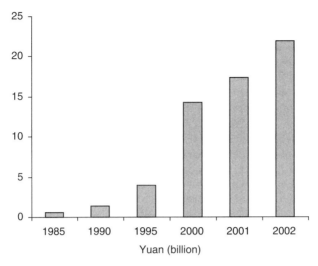

Figure 5.2 Increase of R&D expenditure in universities (1985–2002) (billion Yuan)
Source: MOE, 2002.

degree programs. With over 70 percent of the total research fund contributions allocated, 96 percent of national laboratories and engineering centers have been established, which stands to say that they are the main forces of scientific research work. In these universities there are usually graduate schools, and the ratio of undergraduates to postgraduates exceeds 1:4.

The Ministry of Education in China has not adopted the category of "Research University" yet. Generally, the research university, as a type of university, should have objective standards of evaluation. Combined with the general situation at the international level, and in accordance with the comments of Zhao Xin-Ping (2003), current deputy minister of Education in charge of scientific and technological work in universities, the evaluating indicators to identify research-oriented university includes the following:

1. Sufficient scientific and technological research funds should be available, and the longitudinal research funds coming from different levels of governments should not be lower than the horizontal scientific and technological development funds coming from industries and the private sectors. In the meantime, the contribution of the scientific and technological funds to the operation of universities can be equivalent to the education appropriation granted to them by the government. This contribution refers to the part of the fund that has been transformed into school-fixed assets and staff costs, generally about 30 percent of the total fund.
2. Teachers' scientific and technological workloads should be equivalent to their teaching ones, or the number of teachers whose work is converted to teaching and research should be similar.
3. The ratio of undergraduates to postgraduates should be about 2:1.

4. Innovative scientific and technological achievements that are adequate in number and internationally competitive should be continuously created. As a matter of fact, these can only be regarded as some denotative indicators. For a research university, the conception and idea of running a school is more important. It is necessary to have the concept of setting up a real high-level research university, the idea of having its own superiority and features, and the establishment of operational mechanism of veritable research universities.

China has identified about 20 research-oriented universities, including schools of "the first echelon" and "the second echelon" determined upon the implementation of Project 985 by the Ministry of Education.

Along with the development of scientific research work in universities, the following two policy debates remain heated.

First is the relationship between scientific research and teaching in universities. How to deal with the relationship between teaching and scientific research in universities is a hot issue discussed frequently. The focus of the debate includes: Is scientific research a means of training people or an essential task of the universities? Is there any conflict between scientific research and teaching in universities? Do all of the universities need to do and have the making of scientific research?

With the development of "knowledge-based economy," scientific research in universities is playing an increasingly important role in personnel training, knowledge creation, and social service. The trend that scientific and technological activities have gradually become one of the important ways of personnel training in high-level universities suggests that the society has put forward new requirements for personnel qualifications. With the increase in the number of postgraduates in universities, more than ever, the training of these students cannot do without scientific research activities. In addition, even for the education of undergraduates, due to the rapid development of knowledge, the diverse purposes for enrollment, the reform of problem-centered teaching methodology, and the continuous improvement of teachers' levels, cannot do without scientific research work. However, for certain teachers, a lot of scientific and technological activities will indeed have influence on his/her teaching. Since teachers are more attracted by scientific research, they are usually inclined to *focus* on scientific research and *neglect* teaching. The vitality of scientific research in schools does not naturally improve the quality of teaching. The conflict between teaching and scientific research is more outstanding in universities, particularly for the developing countries, for the following reasons:

1. Compared with the output of scientific research in universities, the country's input to them is relatively small. In 2000, the R&D expenses of state-owned scientific research institutes with independent accounting was 25.82 billion RMB with an increase of 8.3 percent compared with 2003, covering 28.8 percent of the total domestic R&D expenses; the expense in universities was 7.66 billion RMB with an increase of 20.9 percent, covering 8.6 percent of the total; the expenses of different types of enterprises was 54.06 billion RMB with an increase of 23.9 percent, covering 60.3 percent of the total; other expenses was 2.06 billion RMB, covering 2.3 percent of the total. About half of the

scientific research fund in universities comes from the government, while the other half is collected from enterprises (Bulletin, 2000).
2. It is difficult to collect "free funds" for research. The departments concerned mainly introduce project-competitive modes to support scientific research. This marks the tendency of reform; however, not only will the fact that all the projects should obtain funds by means of competition make teachers' work more burdensome, but also this kind of competitive culture will also have an impact on some traditional jobs and the openness of teaching activities.
3. Compared with scientific research institutes, universities have special advantages such as multidisciplinary and interdisciplinary research, concentrated talents, much new blood, and "free and loose" academic environment. The advantage of universities is based on personnel training and basic scientific research; however, in a developing country, the input to basic research is small, outstanding talents cannot be maintained, and the research base is relatively instable, so it will take time to gradually improve the mechanism and research environment. Considering the effect and urgent need of the country, universities are required to carry out scientific research of various kinds, including cooperation with enterprises. As a result, the majority of the scientific research funds of universities is unstable, although research results must be produced in a very short period of time.

In a developing country such as China, a great deal of scientific research work at the current stage falls on applied research and technology transformation. Take the year 2000 as an example (Bulletin, 2000): of the total expense 89.6 billion RMB of Research and Development (R&D), the expenses of basic research was 4.67 billion RMB, covering about 5.2 percent; the expenses of applied research was 15.21 billion RMB, covering about 17 percent; and the expenses of experiment and development was 69.72 billion RMB, covering about 77.8 percent. Although the expense of basic research was increased by 1.22 billion RMB with a rate of 35.3 percent, the funding of basic research is usually small.

Universities have experienced some difficulties in fields such as applied research and technology transformation: (i) There is difficulty in reorganizing the work team. It takes great efforts to organize the research fellows of different teaching and research sections of the same department, or of different departments, to work jointly on a bigger scientific research project, and the team is easy to dissolve; (ii) Sharing resources is another difficulty. Repeated investment occurs in different universities, and the utilization ratio of facilities is low; (iii) Universities are not familiar with the transformation of scientific research achievements and have no advantages. However, the researchers themselves in China usually are required to do the whole process of technically transforming scientific research achievements, which is necessary to obtain economic support from outside the government fund. The result is the emergence of school-run industries of universities and college and science parks related to universities and colleges (discussed later in this chapter). This increases the complexity of campus management, which influences the balance of teaching and scientific research to some extent. A lot of scientific research is carried out in schools in cooperation with enterprises, sending us a new challenge in the common good of school knowledge and the change of school culture.

At present, scientific research is getting more and more expensive, and not all of the universities can afford it. As a result, it is thought that many universities might make some "inquiry learning" and experimentation, not all of them should do scientific research, especially in the field of natural science. Therefore, the Chinese Ministry of Education has proposed guidelines for the scientific research of universities in the years to come, which can be summarized as follows:

1. *Coexistence of multilevel scientific research.* Focus is on basic research, strengthen technological innovation, promote the technical transfer, and normalize the mechanism and the management of industries and science parks related to universities.
2. *Guidance to different areas.* Take dots (key laboratories), lines (interschool cooperation centers, etc.), and then the whole (schools or regions) into consideration, improve coordination, strengthen priority areas, and facilitate cooperation and communication between universities and research groups.
3. *Facilitating.* The gradual form of a range of high-level universities with stronger scientific and technological strength and international competitive ability should be established, making due contributions to scientific and technological advances, economic construction, social development, and state security.

For a country with imbalanced development and vast territory such as China, such developments by levels in research activities in universities, and emphasis on applied sciences, are wise and realistic choices.

*Second is the debate on the development of high-tech industries and science parks in universit*ies. High-tech industries and science parks in universities bear different implications, but they belong to the same category in the course of development and the problems to be debated will be similar in substance. In China school-run factories in colleges started in 1958, when mass activities of "Great Leap Forward" were carried out. Factories for training purposes were established in some schools to enable the students to practice "on-the-job." These factories were mainly the training base and processing plant of teaching and scientific research instruments, and only a few products were supplied to the community. By the beginning of the 1980s, with the in-depth performance of reform and opening up in China—particularly after the State Council issued a decision concerning the institutional reform of economy, science and technology, and education in 1985 focusing on the combination of science and technology, education and economy—a range of universities, headed by the south-eastern University, Central China University of Science and Technology, and the North-Eastern University, personally and successively established science-park-related universities, in different forms, in order to accelerate industrialization and production of new technologies. This (along with the hope of learning from successful examples of the development of science parks related to universities and high and new technology industrial parks at the international level) resulted in the boom of the novelty of the "Science Park," in which the industrial and science parks are directly initiated by the universities instead of the community. However, limited by the awareness and environment at that time, the first attempt to set up college industries and science parks in China failed, but valuable experience has been accumulated.

In 1991, Deng Xiaoping, the general designer of Chinese reform and "opening up," wrote an inscription for the High-Technology Scheme, China: "to develop high technology and to realize industrialization." From then onward universities made a second initiation of setting up high-tech industries and science parks related to universities. Universities such as North-Eastern University, Harbin Industrial University, Peking University, Tsinghua University, and Shanghai Jiaotong University began to march toward hi-tech industries and establish science parks related to universities after this period. The establishment of science parks and the initiation of high-tech industries owned or run by higher education institutes obtained the national support of many sides and played a positive role in the industrialization of high and new technologies and the development of the national economy. A lot of influential hi-tech enterprises and business groups were established such as Tsinghua Tongfang Co., Ltd., Neu-alpine Software Co., Ltd., Founder Electronics Co., Ltd., Tsinghua Unisplendour Group, Peking University Resource Group, Tianjin Tiancai Co., Ltd., Shanghai Jiaoda Only Co., Ltd., and Wuhan Central China Numerical Control Co., Ltd.

In August 1999, the Central Committee of the CPC and the State Council held a national conference on technology innovation and issued the decision of the Central Committee of the CPC and the State Council concerning the "Reinforcement of Technology Innovation, the Development of High Technology and the Realization of Industrialization." It stated expressly,

> Universities should give full play to their own advantages of talents, technologies and information, encourage teachers and researchers to enter the development zone of high and new technological industry and work on the commercialization and industrialization of scientific and technological achievements. They should support the development of science parks related to universities, train a range of high and new technological enterprises and business groups with the intensity of knowledge and intellectuals and the advantage of market competition, making closer co-operation between industry, university and research. (MOE, 1998)

The Ministry of Education (MOE) and Ministry of Science and Technology (MOST) jointly facilitated the science parks related to universities at the national level, first selecting the ones (Bulletin, 2000) in Tsinghua University and Peking University as pilot enterprises. One of the important features of the construction of science parks related to universities is the joint effort among (i) government; (ii) industry; (iii) university; (iv) research; and (v) financial capital. Instead of an action carried out by the individual university, the new action of construction of science parks that would be related to universities received great support from the central and local governments in aspects such as infrastructure, capital, and policy, as well as from enterprises and financial communities. Furthermore, the construction of science parks also tries to follow the guidelines of market orientation and is operated in accordance with modern enterprise system, so most of the industrial and science parks exceed the limit of universities. A lot of domestic and foreign corporations are attracted to settle down in the parks and collaborate with universities. At present, science parks related to universities that are completed and some that are still under construction amount to 44, backed by 104 universities.

Through their painstaking efforts, faculties and scientific research personnel in universities have overcome numerous difficulties and blazed a new trail of developing the high and new technology industry with university-industry-research integration. This is a kind of pioneering undertaking—with Chinese characteristics. The increasing scientific research capability and the rapidly developing scientific and technical industry in Chinese universities have attracted the attention of the whole society. The development of high- and new technological enterprises and science parks has achieved some positive effects:

1. *The enterprises of universities maintain a sound momentum of development* on an increasing scale. Nearly 40 such enterprises of science and technology have been listed in Shanghai, Shenzhen, and Hong Kong stock markets directly or through reverse merger. A number of such enterprises have become renowned at home and abroad, with a very good reputation in society. They are playing an important role in promoting the growth of the national economy and enhancing the adjustment and upgrading of industrial structure. In 2003, the income of university enterprises totalled 82.667 billion RMB (Bulletin, 2004).
2. *They have spurred the development of regional economy* helping to solve many major problems in economic development. For example, the technology of the "efficient filling tower" developed by Tianjin University has been applied to more than 1,000 large- and medium-sized enterprises throughout the country, with the economic benefit amounting to over 500 million RMB.
3. *They have strengthened scientific and technical cooperation* between universities and domestic and overseas enterprises. According to statistics (Ministry of Science and Technology, 2004), by the end of October 2002, the university sci-tech parks had attracted a total social investment of 29.7 billion RMB. The area of incubation parks put into use had amounted to 2.27 million square meters; the number of various research and development institutions to over 1,200 and, enterprises that had moved to 5,500 in number. Nearly 2,300 enterprises were being established and more than 920 had risen up, with 29 of them being listed on the stock market. Totally, 1,860 findings of scientific research above the provincial level had been achieved, with 1,929 having obtained patents and 4,116 new products having been developed. Meanwhile, over 1,300 people who had furthered their study overseas had come back to open their own businesses, providing about 100,000 job opportunities.
4. *They have cultivated qualified all-round talents* with professional knowledge in science and technology and experience in operation and management. The development of sci-tech industry and science parks in universities can help to cultivate talents in two respects. On the one hand, such enterprises and parks provide the environment and platform for business openings, so that people can become experienced in practice. On the other hand, some training courses are provided according to the demand of the market and business, which is beneficial to the cultivation of urgently needed business talents who are specialized in science and technology.
5. *They have promoted the reform of universities.* The development of university science parks has promoted the transfer of scientific findings and the

industrialization of high and new technology. It also enhanced reform of science and technology structure and personnel and allotment mechanism. Universities have adopted humanistic incentive policies to encourage scientific and technological talents to open up businesses in science parks. For example, candidates who have, bachelor's master's, and doctoral degrees are approved to retain their status as students while starting their high and new technology businesses with patented technology or scientific research achievements. They are also encouraged to do so in their spare time or while suspending their schooling. Teachers are also encouraged to transfer their scientific research achievements into products and have a part-time job in high and new technology enterprises or to start their own businesses. Those who open up their businesses are allowed to have corresponding stock shares of technology and management.

However, there are some problems in the development of enterprises owned or run by universities that are embodied in unfavorable property ownership. Other problems are the university's direct undertaking of operation risks, improper management systems, university's overinterference in enterprise administration, and lack of mechanism of "withdrawal" of investment. As a result of these unfavorable mechanisms, these problems are inevitable in the development of enterprises owned or run by universities and are not relatively easy to solve. In November 2001, the General Office of State Council issued the "Guiding Opinions on Pilot Regulation of Management Mechanism in Enterprises Attached to Peking and Tsinghua universities" (General Office of the State Council, 2001), requiring that the structure of university enterprises should be regulated and follow modern enterprise mechanisms. This "Guiding Opinion" is now under implementation and has achieved good results. For example, an assets management corporation has been established in Tsinghua University for the "exercise management of university enterprises."

The hard problem lies in the obvious shock of these enterprises on traditional campus culture. It remains a hot point of discussion on how to look at these changes. As a developing country, China has its market economy at an early stage of development and its scientific and technological forces as well as management talents concentrated in universities. The development of high and new technology enterprises depends on the support of personnel, environment, and scientific research achievements of universities. There is an imperative demand in society for the establishment of enterprises and science parks related to universities. Universities and their faculties also have the enthusiasm to transfer their scientific research achievements to the industry to raise funds. However, it takes a long time for research achievements to transfer, and there also exist risks; therefore, these enterprises first appear in universities, where *two cultures* of different natures coexist: the *original campus culture and values versus new culture and values of enterprises*, whose conflicts will have some impact on the entire university.

The author holds that the achievements of high-tech industries owned by universities and science parks related to universities should be fully affirmed and that some problems like management regulation can be gradually resolved. The development of Tsinghua University and Peking University has proved this. The change of campus culture is the result of the development of the "knowledge economy" and the "knowledge society." In fields such as medicine, computer software, and industrial design, there is an increasingly unclear dividing line between the activities of basic research and establishment of enterprises. Knowledge and wisdom of human beings have become the most important materials required for production. It is more and more difficult to forbid professors to participate in enterprise activities. Therefore, the change of universities is inevitable. What one should do is to try their best to achieve good results in accordance with the trend of the development of human society. The new "knowledge-intensive society" poses new challenges to the *management* of universities. "To meet these challenges, we cannot depend on anything supernatural but on our own exploration and creation in practice" (Wei, 2000).

Zhou Ji (2004) the newly elected minister of Education in 2004 has stressed several times that we "should further emancipate our minds and adhere to the principle of 'active development and regulated management' to promote modern enterprise mechanism." We should promote the positive and rapid development of university enterprises and sci-tech parks toward a new level and a larger scale. It is justified to believe that this is a feasible road for Chinese universities to take in order to "speed up" their development in the "knowledge society."

References and Works Consulted

Bulletin (2000) "Bulletin of Statistics on Chinese Education in 2000." Department of Planning and Development, Ministry of Education, short version at www.moe.edu.cn (accessed July 31, 2005) (in Chinese).

—— (2004) "Enterprises Related to Universities in 2003." Cited from International Forum on Science Parks related to Universities, Tianjin, 2004, www.cutech.edu.cn (accessed July 31, 2005).

General Office of the State Council (2001) "Instructions on Standardizing the Management Mechanism of Pilot University-Own Enterprises in Peking University and Tsinghua University." (In Chinese).

Jiang Zhe Ming (1998) "Speech by the President at the 100 Years Anniversary of Peking University." May 4, 1998.

MOE (Ministry of Education), PRC (1993) "The Outline of Educational Reforms and Development in China, PRC." February 1993. Approved by the Central Committee of the CPC and the State Council, July 3, 1993 issued by the State Council of PRC (in Chinese).

—— (1995) "Interim Regulations of the People's Republic of China on Chinese-Foreign Co-Operation in Running Schools." www.moe.edu.cn (accessed July 31, 2005).

—— (1998) "The Action Plan of the Revitalization of China's Education to Meet the Needs of the 21st Century." Published by and approved by the State Council in January 1999 (in Chinese).

—— (1999) *The 50 years of Education of People's Republic of China*. Beijing: Ministry of Education and Higher Education Press.

—— (2002) *Chinese Education at the Turn of the Century*. Beijing: Higher Education Press (in Chinese).

—— (2003a) "Bulletin of Statistics on Chinese Education in 2002." Department of Planning and Development, Beijing (in Chinese), short version at www.moe.edu.cn (accessed July 31, 2005).

—— (2003b) "Bulletin of Statistics on Chinese Education in 2004." Department of Planning and Development, Beijing (in Chinese).

—— (2004) "Bulletin of Statistics on Chinese Education in 2003." Department of Planning and Development, Beijing (in Chinese), short version at www.moe.edu.cn (accessed July 31, 2005).

MOST (Ministry of Science and Technology) (2001) "Bulletin of the Statistics on Chinese R&T in 2000." Beijing (In Chinese).

—— (2004) "Bulletin of Science and Technology Statistics." Ministry of Science and Technology of the People's Republic of China, Beijing (in Chinese).

National Centre for Education and Research (2002) *Green Paper on Education in China: Annual Report on Policies of China's Education*. Beijing: Education Science Publishing House.

Project 211 (1995) "Project 211." Approved by the National Commission of Planning as part of the Ninth Five Years Plan of National Economical and Social Development, www.moe.edu.cn (accessed July 31, 2005).

Wei, Y. (2000) "The Development of the University-Related Science Park in China." Keynote Speech at the Fourth Annual Conference of East Asia Science Park, Hsinchu, Taiwan province of China, October.

Zhao Xin-Ping, (2003) "Report by the Vice Minister of Education." (In Chinese), www.cutech.edu.cn (accessed July 31, 2005).

Zhou Ji (2004) "Plan for the New Break in Reform and Leap Forward in Development, by the Minister of Higher Education." *Xin Hua Wen Zhai* (Chinese journal), 21:102.

Chapter Six
Between the Public and the Private[1]: Indian Academics in Transition

Karuna Chanana

Introduction

Higher education (HE) is in the midst of a fundamental shift in its relationship to society as a result of profound economic and political changes. On the one hand, public support of higher education is in question due to the dominance of market ideology; on the other hand, the social demand for higher education is rising. Higher education is being equated to a marketable commodity and as investment for profit. It is no longer viewed as a public good. The private sector is coming to play a critical role in higher education even in countries where it was, until recently, fully subsidized by the state. The question asked is, Is higher education no longer the social responsibility of the government? This is especially true when it is also being argued that higher education is a keystone to development—be it economic, social, or human. Additionally, since higher education is expected to ensure trained and responsible leadership, it can be said to provide the main thrust in development (Thompson et al., 1977). The universities and the academics are also being perceived as critical to knowledge production and to train the human resources for the restructured global economy. This is putting pressure on the universities to change their goals and functions, which impacts on its traditional functions and also on the academics.

The terminology of discourse has changed—financial support has become "subsidy"—a word lifted from the economic discourse and transplanted into the educational arena. "Brain drain" has become "migration of knowledge workers" and knowledge generation/creation is now "knowledge production." Now higher education is viewed as a nonmerit good that has to be paid for by those who would like to acquire it. The *buzz words* are self-financing, marketization, privatization, industry higher education interface, and so on. Perceptions about the academic profession have changed from one that prepared professionals to one that produces a "globally competitive workforce" (Kelso and Leggett, 1999).

While the higher education system, as a whole, is in a turmoil this chapter focuses on the impact of the redefinition of the role of the government in higher education on the academic profession in India generally, and on the teachers in the colleges and universities specifically. It is based on the understanding that higher education is changing considerably in response to economic reforms of post-1990, which include changes in state policy vis-à-vis higher education, rising social demand for higher education, internationalization of higher education, and also it's shift from a largely publicly funded system to the market and private funds.

The changes across higher education in India are very rapid and continuous with very little documentation on it. Academic profession as an area of investigation has not been researched in India. Most importantly the choice of the theme of this case study has been influenced by the author's own position as a social science teacher in a central[2] university. This chapter will refer mainly to the social sciences in the universities and colleges and occasionally to the specialized institutes of science, technology, etc. In this instance, the public system of higher education includes the private-aided colleges while private institutions, in this chapter, refer to the unaided, self-financing colleges/institutes such as the Deemed universities[3] and private universities.[4]

This chapter

- provides an overview of higher education in India especially in the pre-globalization phase, that is, from 1951 to 1991 and highlights some of the salient features of the academic profession;
- discusses the developments since 1990 when the government of India adopted the policy of economic reform and liberalization;
- highlights the expansion of the private sector, especially the private unaided self-financing institutions that have come up during the last decade or so; and
- looks at some of the issues and problems facing the academics in conditions of marketization and the extent to which the academic profession has undergone transition.

Knowledge, Academic Profession, and the University

The classical organizational structure for universities is that of an institution that is governed by an academic community based on collegiality. Collegiality is based on the concept of the university composed of a community of equal scholars who can manage their own affairs and activities and act as a "clerisy" or a body of scholars giving importance to the collective rather than individual decision-making process (Farnham, 1999).

It is not for the first time that the contribution of state support to the universities has become central to the discourse on higher education. In late eighteenth century, Adam Smith advocated the need for competition and accountability in the universities. He supported the idea of paying the salaries out of the tuition fees paid by the students, rather than out of endowments, so that they could demand and get the kind of education that they had paid for. This would generate competition and encourage the faculty to improve its performance (quoted in Bhushan, 2004). Buchanan and Devletoglou (1970) argue that competition is essential, even if there are no tuition fees, especially competition among universities for faculty appointments to reduce mediocrity.

The centrality of competition versus cooperation within the universities and endowments/state support versus financing through tuition fees have also been the concern of other scholars. For example, Veblen (1993) underscored the autonomy of scholars in the universities and said that cooperation is the essence of quality and excellence in scholarship for which endowments and state support are needed. Veblen was also against comparing the universities to business enterprises and said that "competition is anathema to the pursuit of knowledge." Competition will give rise to mediocrity, while quality of the university depends on the quality of research and the academic reputation of the scholars. Cooperation is the driver of excellence.

The traditional universities have also been an instrument of economic growth, and they were expected to contribute to social and economic progress by training a qualified and adaptable labor force, for example, high-level scientists, professionals, technicians, schoolteachers, future government, civil servants, and business leaders; by generating new knowledge; and by building the capacity to access global knowledge and to adapt new knowledge to local use. The World Bank (WB) takes the view that the state has a responsibility to provide an enabling framework to the traditional educational institutions, which will spur them to innovate and be responsive to the challenge of the knowledge economy. The continued government support to tertiary education is justified on grounds of the external benefits such as health, agriculture, equity issues, and the supportive role of tertiary education in the whole system of education (World Bank, 2002).

Perkin (1969:1f) describes the academic profession as "the key profession of the 20th Century." According to him, universities, through their academic staff, provide "the growth points of new knowledge, the leading shoots of intellectual culture, and the institutionalization of innovation in arts, sciences and technology. Therefore, academics are the creators of new knowledge, critics of conventional academic and epistemological wisdoms." However, as professionals they also have their occupational and material interests as well as disciplinary needs. These include the freedom and space to teach; to study their academic subjects without external interference; the right to participate in the decisions relating to curriculum and research agenda; the right to participate in the running of the institutions; security of tenure; and satisfactory terms and conditions of employment (Farnham, 1999:3).

Thus, the traditional expectation from a teacher has been that he/she who pushes the frontiers of knowledge will be the best instructor. Peer review and publications resulting from research have been the most common method of measuring the output and quality of research and also evaluating the performance of university teachers. According to Farnham the quality of teachers, how they are hired, rewarded, utilized, and motivated have been viewed as critical to the efficiency and effectiveness of the higher education system (1999:x). Eustace (1987) mentions five criteria that define the collegial approach to academic governance. These are (i) equality; (ii) democracy; (iii) self-validation; (iv) absence of nonscholars; and (v) autonomy from society, especially freedom from political and state intervention.

Trow (1997:26) divides the life of higher education system into public and private. Public life refers to organization, governance, and finance, while private life is what is experienced in the classrooms, libraries, seminars, and during the teaching-learning process (Farnham, 1999:26). However, the academics are interested not only in curricular and pedagogical issues that fall in the domain of private life but also in the ways in which universities are governed and managed and, therefore, in the public life of the institutions in which they work. Trow (1996) is also aware of the changes in higher education and in the role of the state. However, Trow (1996) expects a shift to take place after the systems have reached a certain level of penetration in the society and, for example, says that mature, mass systems of higher education have over 25 percent of the age cohort entering them; Trow (1996:17) feels that "massification leads to more diversity of institutions, greater institutional autonomy and diversification of support so that the state's contribution to the system falls as higher education shifts to the market."

Kogan (1988:68f.) divides institutions into those that are independent and others that are dependent and defines collegiums as "a minimum organization" where the independent college comes together to admit new members, establish minimum standards, and divide its common resources. The dependent institution is the one whose objectives are not set by the members of the academic profession but by their sponsors. "The academics will still determine how to do it, but ultimately what should be done, and why, will be determined by external forces" (Kogan, 1988:71). Furthermore, under the dependency model, education for its own sake is not the function of higher education. It is to meet the national goals for trained people in the labor market and produce useful and utilitarian knowledge. While Kogan accepts that most institutions have been a combination of the two models, the author also says that "in the past academics rejected or accepted such dependency on their own criteria. They did not have to do what they were told in order to keep in employment. Their substance was not determined by a string of *ad hoc* contracts" (Kogan, 1988:71). Ultimately, academic professionalism depends upon effective, individual professionals who have the space and autonomy within their institutions to teach and to undertake research on the issues that are at the cutting edge of their disciplines. In practice, however, the academics have never enjoyed unfettered freedom in any system of higher education. However, an essential precondition for self-management and governance is independence from financial pressure, which is possible only through what Farnham (1999) calls "benevolent" state grants or private endowments that allow the institutions to set their own goals and formulate their teaching and research programs. "The basic intellectual driving force behind the collegial model is the objective pursuit of knowledge by individual scholars" (Farnham, 1999:19). These prerequisites seem to be missing from the private institutions and are also rapidly disappearing from the state[5] universities.

One of the most important functions of higher education is to create, apply, and disseminate knowledge. Knowledge is also its most important resource. According to Lauden and Lauden (1999) higher education is essentially about the creation, transformation, and transmission of knowledge, social and economic progress depend on its application and advancement. Knowledge is created/produced, nurtured, applied, and disseminated by the individual faculty and researchers through their teaching, research, published and unpublished sources (curriculum and research reports), and so on. Thus, the most important resource in the university system for knowledge economy is the talents and expertise of the faculty that can promote knowledge innovation (Thakur and Thakur, 2005).

However, the importance and definition of knowledge in the global world has changed radically. Scholars have identified two types of knowledge, namely, "explicit" and "tacit" knowledge. While explicit knowledge is documented and captured through print and ICT, tacit knowledge is intangible and cannot be easily expressed, for example, hunches and curiosity that lead to research (Nonaka and Takeuchi, 1995). However, the emergence of new types of institutions, forms of competition, and technological innovations are pushing traditional institutions to change their mode of operation and delivery. According to Norris et al. (2003) so far knowledge has been a cottage industry, and it has to change in the face of the networked world. The colleges and universities have to move from a culture of "knowledge hoarding" to one of "knowledge sharing."

Universities and Globalization

Globalization covers a wide variety of changes, namely, technological, economic, cultural, social, and political, all of which transcend the boundaries of the nation-states. The forces unleashed by it impact universities all over the world. According to the World Bank (2002) universities and higher education institutions face new trends in the borderless global environment that affect the very purpose of tertiary education systems, modes of delivery, and organizational patterns due to the IT revolution. One of the most important dimensions of change is the emphasis on knowledge as the main driver of growth and the IT information and communication revolution. It is argued that knowledge generation and accumulation, dissemination and application are critical factors in economic development and are essential for a country to have competitive advantage in the global economy. The World Bank (WB) also emphasizes the role of higher education in democratic societies for social and economic progress and its role as an important global public good. Furthermore, it is essential for the creation of intellectual capacity on which knowledge production and utilization are dependent. It also warns of the danger of a growing digital divide across and within nations due to the availability or nonavailability of technological know-how, preparedness, and usage.

In India too globalization has wide ramifications. India's problems are manifold, namely, its inability to expand tertiary education coverage after the 1980s, inequalities of access and outcomes, problems of educational quality and relevance, rigid bureaucratic procedures of governance structures, and management practices. However, this chapter will focus on four visible dimensions of the impact of globalization on higher education: (i) limited state funding; (ii) privatization; (iii) self-financing by students; and (iv) substitution of tenured/permanent positions of teachers with contracts and low pay, as these are pertinent for the academic profession at present because of their direct impact.

While higher education in most of the developed countries became available on a mass scale due to full public subsidy, this process is being halted midway in the developing countries. When the enrollment in higher education is less than 10 percent of the relevant age group, restructuring of economy due to globalization and economic liberalization has overpowered the expansion of higher education. The corporate influence on higher education is more marked in the developed countries (Gibbs, 2001). The universities are becoming or expected to become extension centers of the high-tech industries and the corporate system. The rules of intellectual property rights are determining the parameters of research in the university and are pushing researchers to work for profit.

Two simultaneous trends are taking place in higher education. On the one hand, the reduction of, or, worse, the freezing of, state support from public universities is pushing these universities toward raising funds from the individual student. On the other hand, private universities and institutes of higher education are being established in countries even where there were no private universities. Scholars are differentiating between education companies and nonprofit higher education (Garrett, 2003:9). What is emerging is a competitive higher education marketplace where "a higher share of the state resources is the most sought after prize for most" (Brennan et al., 1999:4). A differentiation is being made between research universities and modern (teaching) universities (Farnham, 1999), the former referring to the traditional university and the latter referring to the university that responds to market demands.[6] Universities have to make choices regarding knowledge and curricula, access and participation; teaching and learning; and decision making and accountability (Brennan et al., 1999:5).

The Indian Context

In the context of an emerging democratic polity, the political elite, the social reformers, and intellectuals agreed, at the advent of Independence in 1947, that higher education had to be an effective instrument to promote economic growth, social change, and social and economic mobility. A widespread belief in India is the right of access to higher education and has been manifest in the pattern of government funding. Therefore, the state had to ensure that access to education was easily available and inexpensive, if not free, to enable the masses especially the disadvantaged and the first generation learners to enter the system. Therefore, the government took direct responsibility for the provision of higher education instead of shifting the responsibility to the individual and to the private initiative.

At that time too the relationship between education and economy was perceived as quite critical and close and, therefore, science and technology were given a special place in higher education. Institutions of international standards were established in engineering such as the Indian Institutes of Technology (IITs). Simultaneously, general education too was not neglected. Existing universities were provided support along with newly established universities. The emphasis was on promoting excellence in universities generally and to support research in specialized institutions along with expanding undergraduate and graduate education in the colleges. In order to retain the autonomy of universities, the University Grants Commission (UGC) was established to provide financial support and to ensure quality and set standards. It was expected to play a critical role in setting standards for quality of teaching (including student teacher/ratio, teaching workload, etc.), recruitment and promotion criteria, and salaries of teachers.

Economic reforms led to the restructuring of the economic institutions and the education system. The government of India began to change its thinking about financial support to higher education after 1991 when economic reforms were pushed by the international funding agencies by "accepting the rightists' project of economic competitiveness and rationalization" (Apple, 1997:595). Now higher education is being viewed as a facilitator of national competitiveness, economic growth and development, wealth generation, producer of knowledge workers, major source of business knowledge and, users of technological innovations (Farnham, 1999:x). This has had an impact on all the aspects of the higher education system and changed its goals and purposes as well. The economic value of higher education has gained primacy at the cost of social benefits. This overshadows the earlier view that the government had a social responsibility to provide free or inexpensive and quality education to all because it was an instrument of economic growth that goes hand in hand with social uplift and mobility for the disadvantaged groups. Therefore, there has been a shift from the earlier policy of full financial support to higher education to allow the market forces to give direction to it.

Since the early 1990s one has been hearing, as a part of the public discourse on Indian economy, that

- higher education should be self-financing;
- public funding to higher education should be reduced; and
- higher education and industry interface are desirable and should be promoted.

In sum, higher education should raise funds for its survival and development. In addition to the international pressure for economic reform, the increasing demand for higher education and for specific academic programs and courses also became push factors. Both the central and the provincial governments relaxed their regulations and allowed the entry of the private sector more freely. This was also necessitated because the existing system had become too large and dysfunctional. The dysfunction had not only to do with size but also with popular democratic politics whereby reforms, financial or academic, were not possible due to the political pressures. Quality had become a martyr in the process.[7] The international pressure came at a time when the disillusionment with the public sector and the socialist model was almost complete and the belief that "public is . . . the center of all evil; private is the center of all that is good" (Apple, 1997:596; Tilak, 1999:113) was widespread. This is not to deny that there were no quality institutions and quality teachers in the higher education system and that quality is being ensured in the new private institutions that are mushrooming overnight.

Furthermore, the increase in social demand for higher education and the inability of the governments to meet the financial costs has pushed higher education toward privatization. India has had a long history of private participation in education under colonial rule when the goal was to educate the people and also to become self-reliant. The Indian social reformers and political leaders played a seminal role in the establishment of universities such as the Banaras Hindu University at Varanasi in Uttar Pradesh, the Aligarh Muslim University at Aligarh, also in Uttar Pradesh. However, private institutions were generally established by religious and charitable organizations and by private trusts and foundations that were set up to be the reformers and leaders. In the posteconomic liberalization phase, on the other hand, private institutions are started by for-profit private agencies. Even the charitable organizations of preindependence days have changed color and have started for-profit colleges on self-financing basis. Such for-profit private institutions have been established in large numbers. Their sole source of income is the tuition fees paid by the students.

Thus, a combination of less regulated environment, competition in the higher education sector for public and private funds, and accountability has brought about several changes. This has helped in the diversification of funds. Tuition fees have been introduced where none existed. Student loans are being made available either through the banks or directly by the universities to neutralize the higher individual cost of education.[8] Private colleges and universities are introducing courses that have a demand in the market. Public universities have also been allowed to generate income from the self-financing courses. Additionally, faculties are expected to raise funds through projects, patents, and consultancy.

Indian Higher Education System

Until recently, the central and the state governments have been the main providers of higher education since higher education is a concurrent subject and comes under both the central/federal and the state/provincial governments. There are some central universities while the majority are state universities.[9] None of the universities were private till recently although there were privately aided colleges affiliated to the universities. These colleges receive financial aid from the state government varying from 90 to 95 percent (Tilak, 1999:121). For all practical purposes, there is hardly any difference between the government and the privately aided institutions and, therefore, they are included in the category of government/public institutions.

The best institutes in management, engineering/technology, and medical education were and still are in the public sector established by the federal government. These are Indian Institutes of Management (IIMs), Indian Institutes of Technology (IITs), and the All India Institute of Medical Sciences (AIIMS) in New Delhi.[10] Four IIMs and five IITs were established in different parts of India. In addition, there were regional colleges of engineering. Some institutes of engineering and management were also established by reputed trusts and foundations. But they were too few until very recently.

The Indian higher education system has expanded phenomenally in the past 5 decades of independence from 1951 to 1991 but more so in the first 3 decades. In 2002–2003, there were 300 universities that included 19 central universities, 183 state universities, 71 institutions that have been granted the status of deemed universities, and 13 institutions of national importance. Universities are of two kinds, namely affiliated and unitary. A majority of Indian universities fall in the former category, that is, colleges are affiliated to them. While colleges undertake teaching, the curriculum, examination, and evaluation are organized by the university departments that also undertake postgraduate teaching and research. The unitary universities have mainly departments of postgraduate teaching and research. In the same year, there were 15,343 affiliated colleges. Out of these, 1,650 (10.75 percent) were exclusively for women students. The enrollment of students was 9,227,833 out of which the number of women students was 3,695,964 (40 percent). Even though the number of students in higher education is more than 9 million[11], only about 8 percent of the relevant age group enters higher education. This is in contrast to the developed countries where more than that 40 percent enter higher education (Tilak, 1999).

Nearly 90 percent of the students are enrolled in the undergraduate programs in the arts, sciences, and commerce faculties in the affiliated colleges. Additionally, 66.2 percent of graduate students are enrolled in the affiliated colleges. On the other hand, 91.15 percent research students are enrolled in the university departments (UGC, 2003:125).The number of teachers has decreased from 457,000 in 2000–2001 to 436,000 in 2002–2003. Of these, 82 percent were teaching in the colleges in comparison to 18 percent in the universities. The proportion of professors in the colleges and universities was 6.63 percent and 22.2 percent respectively; 21.12 percent and 31.53 percent readers/associate professors respectively; and 66.70 percent and 45.75 percent senior lecturers/lecturers respectively (UGC, 2002:126). The remaining teachers hold the positions of tutors and demonstrators.

India's Contribution to Knowledge Production

It was reported in the newspapers that India ranks second (38,195) to the United States of America (239,000) in terms of distribution of certified professionals in nine major categories (Rajagopalan, 2003:13). Chaudhuri (2005) mentioned that *The New Scientist* has dedicated its February 2005 issue entitled, "The Next Knowledge Superpower" to Indian technology from software to satellites to pharmaceuticals. He also reports from the Global Skills IQ 2003 Report brought out by the United States of America firm Brainbench. It refers to the erosion of the status of the United States of America as the Technological (Tech) Superpower of the world. It projects India as the next Tech Superpower. According to *The New Scientist* (2005) less Americans are getting science and technical degrees and the United States of America is spending less on civilian R&D though it is still bigger than the expenditure of the next five countries put together. The rate of growth of USA science and engineering publications was 13 percent between 1988 and 2001 while it grew by 25 percent in India. However, the United States of America still produced 20 times more than India in 2001 (Chaudhuri, 2005:1).

The number of Indian immigrants with tertiary education in the United States of America is 22,500,000 (World Bank, 2002). The "New Scientist India" mentions two trends that push India onto the fast track for knowledge production and contribution. The first is referred to as "brain circulation". Since the Americans lost interest in laboratory careers, the United States of America has been compensating that by bringing foreigners from overseas. Now the Indians and Chinese are going back to their countries where they are getting better opportunities. The second trend relates to the globalization of technology due to which USA firms are outsourcing R&D to foreign countries. India is quite high on the list of business process outsourcing (BPO). *The New Scientist* (2005) estimates that more than 100 IT and science-based firms have located R&D laboratories in India. It gives the example of the General Electric (GE) Laboratory in Bangalore, which is well known for its material sciences division. Another example is of a firm that has recently got a contract to commercialize the USA nanotech drug—delivery patent (Chaudhuri, 2005:13).

It would appear that according to the Global Information Technology Report (Dutta et al., 2004) India ranks forty-fifth out of 102 countries on the Networked Readiness Index (NRI). India was 37 out of 82 countries in 2002–2003. The NRI index consists of three components: (i) environment; (ii) readiness; and (iii) usage. The Indian government ranks well on prioritization of ICT, but 71 on procurement; 31 on readiness of ICT, and 26 on its usage. India ranks third (it was second in 2003) in availability of scientists and engineers, it ranks twentieth in the quality of scientific research institutions. As the number of qualified and trained professionals is migrating, its rank on brain drain has slipped from 54 in 2003 to 65 in 2004. This indicates that the heavy investment in science and technology does not benefit the Indian economy (Dutta et al., 2004).

Apart from the migration of knowledge workers from India, the latest is the internationalization of education that is attracting Indian students to higher education in the developed countries of the world with the United States of America being on the top, in spite of a slow down after 9.11. Indian students are being offered financial incentives. Beginning the year 2005, the European Commission has set aside 500 million Euros to grant scholarships/fellowships to 900 Indian students over the next 3 years. There is competition to attract Indian students. With international schools opening in India and the Indian schools offering international baccalaureate, students are prepared to go abroad for undergraduate education. This will have an impact on the quality of higher education in India and the academic profession. *Will it be able to retain its advantage?* Although its rank is 84 in public expenditure on education (per capita in 2000); 72 in tertiary enrollment (gross percent in 2001); and 80 in scientists and engineers in our R&D (per 1,000 inhabitants in 2000), it is fairly high (14) in quality of mathematics and science education and in the quality (8) of business schools.

Support for Professional Development of Faculty

Provisions have been made to ensure that teachers could upgrade their qualifications, undertake research, have access to best libraries; are able to participate in conferences and seminars at home and abroad; and are able publish their doctoral theses, and so on. Although the funds set aside for the purpose may seem small in comparison to the large number of teachers, it signifies the thinking of the government about the role and function of the academic in higher education. Provincial governments also provide funds for some of these activities. These are mentioned here to give an idea of how the professional development of a teacher in higher education was visualized and planned.

For example, since 1963 Special Assistance Programmes have been introduced to promote excellence in science, engineering, and technology, and humanities and social sciences. So far 144 departments in humanities and social sciences and 258 in science, engineering, and technology are receiving this assistance. Another exclusive scheme provides assistance to departments in science and technology for acquiring state-of-the-art equipment of international standards for postgraduate education and research.[12] The number of such departments is 217 in 58 universities. To promote quality along with sharing of resources University Science Instrumentation Centers (USIC renamed as Instruments Maintenance Activity) have also been set up. The equipment in the 74 centers is shared among universities. Introduction of computer applications and information technology in the universities has also been accompanied by financial support. In the Tenth Plan provision has been made to financially support computer connectivity in colleges through the UGC-Network Resource Centres.[13] Moreover, research, fieldwork, and travel to foreign countries are available under the Area Studies Programme of the UGC. There are 20 area study centers, such as the West European, American, African, and Mid-Eastern, in some universities.

Quality concerns have been addressed by the UGC, which is the apex organization established to monitor standards and to disburse funds. Recently the UGC decided to identify universities and colleges of *potential excellence* and give funds in addition to the regular annual grants. It set aside INR3 million for 10 universities and INR5 million for 300 colleges of excellence. So far 5 universities have been identified and given INR300,000,000 each. In addition, 150 colleges have been given a onetime grant of INR10,000,000 each.[14] It is also reaching out to colleges and is providing equipment and Internet connectivity for networking and classroom usage. Network Resource Centres are being set up in 6,000 colleges. In this scheme, the specific focus is on the colleges located in the rural and backward areas. All colleges are eligible, and each college will be given INR500,000 to 1,000,000 for this purpose. Colleges in the rural areas will get a onetime additional grant of INR3,000,000 for improvement in infrastructure such as library, laboratories, and equipment. For networking and connectivity three areas have been identified: (i) science; (ii) liberal arts and humanities; and (iii) social sciences. UGC is also pushing the introduction of e-learning in all the colleges and to train the teachers through its 50 Educational Multimedia Research Centres across the country. Later these will also ultimately be connected to EDUSAT, a satellite launched by the Indian Space Research Organisation (ISRO) in August 2004.[15]

For promoting standardization in the quality of teaching and research, the National Eligibility Test (NET) had been introduced for recruitment of teachers in universities and colleges and also for receiving research scholarships for pursuing doctoral studies. Some states have also introduced state-level tests for recruitment of teachers. The UGC also provides financial aid to teachers for completing their MPhil and PhD under the Faculty Improvement Programme. During the Ninth Plan 2,800 Teacher Fellows were financially assisted in their research. In 1997–1998 research awards were started, and by 2002–2003 95 teachers (74 men, 21 women) had been selected for these awards. In order to improve the quality of teaching and research in the colleges and universities that are not situated in the metropolitan cities or do not have access to the best talent in teaching and the library resources the UGC also has a scheme of visiting professors and the teachers. Under this scheme teachers who are well established in their fields are selected to visit a number of institutions across India for a few months. There are also national lectureships and adjunct professorships.

Academic Staff Colleges have been established for the professional development of university teachers and for updating their knowledge base. There are 51 academic staff colleges and 96 universities and specialized institutions that offer short-term refresher and orientation courses to teachers in higher education. Training has been made mandatory for teachers for promotion. Up until 2001–2002, funds were given for 225 orientation and 983 refresher courses, and 217,000 teachers have participated in these programs.

In addition, faculty members can apply for fellowships from the Indian Council of Social Science Research (ICSSR) during their career in order to pursue research. They are entitled to get academic leave from the university or the college to utilize the fellowships. Even after retirement the faculty members can receive Emeritus Fellowships from the UGC. There are 100 fellowships under the scheme. The ICSSR also awards Senior Fellowships to working teachers for research and National Fellowships to outstanding scholars in service for those who have retired up to the age of 70 years.

Financial support is provided to university teachers for presenting papers in international conferences by the UGC,[16] ICSSR, Indian Council of Historical Research (ICHR), Council of Scientific and Industrial Research (CSIR), Department of Science and Technology (DST), and so on. Fellowships, publication subsidy, funds for research projects and for organizing seminars, and so on are also available to faculty from the same agencies and organizations. In addition the Planning Commission ministries and their departments, for example education, social welfare, women and child development also provide funds for applied research.

In addition to this, there are bilateral and multilateral Cultural Exchange Agreements/Programmes (CEP) with foreign countries under which faculty members can go abroad for doctoral and postdoctoral research. Travel grants are also available under these programs for collection of source material for research. UGC operates bilateral CEPs with 45 countries such as the German Academic Exchange Service (DAAD), the Shastri Indo Canadian Institute, and the Commonwealth Academic Staff Fellowships.[17] In addition, the universities can use the unassigned grants given to them for participation of their faculty members in international conferences.

Status and Working Conditions of Faculty

In India, the basic responsibilities/duties of the academics are teaching and research while extension/service to society remains marginal. Thus, it is possible to talk of only teaching and research as the two main expected duties of the teachers. The binary division referred to above is also reflected in the division of teaching and research, that is, most of the college teachers undertake undergraduate teaching while those in the universities undertake both graduate teaching and research. In fact, most of the colleges do not provide any facilities or working environment for teachers to pursue research nor do the teachers perceive the professional need for research.[18] Most of them do not devote their full-time to teaching. This is in spite of some of the very high quality colleges that have promoted excellent teaching and research among their faculty. Research, on the other hand, takes place in the universities and in the specialized institutions of higher education as well as institutions set up for the explicit purpose of promoting research.

Following the UK system, the higher education system classifies the faculty into three categories: professors, readers, and lecturers. Some universities, which were established later, follow the USA system and use the designations of professors, associate professors, and assistant professors. Institutions such as the IITs add the categories of lecturer and assistant lecturers below the three higher categories. There are fixed scales and salaries according to the designation and specified criteria.[19] Salaries are public knowledge. The gross salary of a lecturer may vary from INR16,000 per month for a lecturer to INR30,000 for a professor. Increments are given annually—the amount of increment is insignificant and depends on the salary scale. "Dearness allowance" is given to offset the rise in the cost of living. The salary scales are revised upward periodically, approximately every 10 years, by the Ministry of Human Resources and Development (MHRD), government of India, and the UGC for the teachers in the central universities and institutions. The provincial governments also follow suit sooner or later, although there are marginal variations. The recruitment and promotion criteria for each category, namely, professor, reader/associate professor, lecturer/assistant professor are fixed along with the salary scale by the MHRD. All the posts are advertised in leading newspapers for wide publicity, and the criteria for selection are made public. In this respect too most of the provincial governments generally follow the criteria set by the MHRD although they may take longer to upgrade the salaries.

One of the important recruitment criteria is reservation on the basis of a ethnic quota system based on caste and tribe in addition to those who belong to the officially identified categories of Other Backward Classes (Chanana, 1993). The Indian Constitution provided for reservation of 22.5 percent positions for the Scheduled Castes and Scheduled Tribes for recruitment to the first level of teaching positions, namely, lecturers. Political pressures are building up to allow the same criteria for promotion.

So far as the salary and tenure are concerned, until recently, all the teachers were protected by full-time positions and regular salary scales. The probation period varies from 1 to 2 years, which is only a formality. After this the teachers are confirmed in their jobs, that is, they become permanent or hold a tenured position that entitles them to pensions and provident funds. Medical facilities, car loans, sabbatical/study leave, and other facilities are provided by the universities that are run by the central government but they vary from one state university to another. Some universities and colleges that have a few houses on campus provide accommodation to a few teachers for a nominal rent. Once recruited for permanent positions, therefore, the teachers possess a job for life or what is referred to as the "iron rice bowl" (Chen, 2002:112), and it is very difficult to terminate their services even if they are found derelict in their duties. There has been no difference in the salaries of those who are rated as good teachers and researchers. Such posts have been dwindling since the 1990s in the state universities.

The age of retirement varies from 58 to 65 years. The age difference has also led to the exodus of teachers from universities. For example, one of the state universities has lowered the age of retirement from 60 to 58 years, and, therefore, this is a reason to look for jobs in other universities. Those with good professional standing are moving out.

Teachers are also entitled to postretirement benefits such as provident funds and pensions. However, retirement benefits are not sufficient to meet the cost of living in the postretirement period. Even the retiring full-time professors from a central university will require sources of income other than the pension.

The right of teachers to unionize became very crucial and was recognized by the government as critical to the nonexploitation of teachers. Teachers were and still are characterized by a "hypocritical psyche."[20] The unions received direct support from all the political parties and were in a position to put pressure on the vice-chancellors to succumb to their unreasonable demands that resulted in the lack of accountability and nonperformance among the teachers and the other sections of the university community. In fact, the whole public sector in the economy was marked by these features. The politicians not only protected the teachers from performing but also came to play a very crucial role in the recruitment and promotion of teachers. Therefore, political ideology and political connections became more important than competence in teaching, publications and research. Another fall out of democratic politics was that the politicians were the first to set up the profit-making private unaided self-financing colleges. This development is discussed later.

The linking of democratic politics also led to the erosion of the autonomy of the universities, especially of the state universities. In the course of time some of the provinces introduced laws to curb the powers of vice-chancellors. In fact, in some cases the appointment of the vice-chancellors was taken out of the Departments/Ministries of Education and was put under the chief minister. Thus, universities became virtual extensions of the governments. The impact of this for teachers was negative as well as positive. In Bihar, one of the most backward provinces in India, the chief minister introduced automatic promotion for the teachers sometime in the early 1980s. As a result most of the teachers became professors without having the required qualifications, experience, and publications. A few other provinces also followed suit. The negative impact was that once the vice-chancellors became the direct nominees of the politicians their academic credentials became questionable and so was their ability to appoint and promote good teachers.

Nevertheless, the teachers' unions have played an important role in negotiating with relevant ministers/bureaucrats in fixing workload; number of working days and vacation; fixing criteria for recruitment and promotion and also the upgradation of salaries from time to time. As a result, teachers' workload and the number of teaching hours have been ridiculously low for the past several years. So far as the salaries are concerned, the central government has had a policy to offset the cost of living by periodically (in a little over a decade) upgrading the salaries of staff in the public sector. The teachers' unions have had to wage a struggle for their salary upgradation. The time gap between the revision/upgradation of salaries and the erosion of the purchasing power is generally too long. As a result, several teachers especially in the colleges would start looking for additional sources of income such as giving private tuitions, coaching, establishing private schools in the names of their spouses, and so on. The male teachers are likely to do this more often. Private tuitions were easy because college teaching was only half-time due to light workload and lack of accountability, and teachers had plenty of time to spare. They were also protected by the trade unions that were not known to raise issues of teachers' accountability and quality of education. Thus, over a period of time teachers' unions have come to stand for extreme protection by the politicians, in the ruling party as well the opposition, to the detriment of teaching standards and quality in the higher education system.

In 1990 when the Indian government decided to introduce economic reforms, the education system was also put on the agenda because it, barring some exceptions, became obsolete, nonperforming, and did not meet the expectations of the country. More specifically the complete indifference of the teachers to the quality of education and their social responsibility contributed to the thinking that something drastic needed to be done to the publicly supported education system.

The paradox of Indian higher education is that even though it is second in terms of the certified professionals and high rate of outmigration of skilled and educated knowledge workers to the United States of America and other developed economies, a majority of the products of HEIs are of doubtful quality. Moreover, there are very good institutions along with a majority of inefficient colleges and universities. It is highly differentiated in quality and size and is like a pyramid with a few exclusive institutions at the top and a huge number of institutions of indifferent quality and of a larger size forming the base. This hierarchy and the stratification are also reflected in the teaching community within and across the institutions. In other words, a few teachers at the top have qualifications of international/national standards while the remaining mass of teachers is at the bottom. Therefore, there is a binary division in the quality of institutions and of teachers.

Changing Environment: Privatization of Higher Education

The changes in the higher education system in India cannot be understood without looking at the global trends. While the period before 1991 or the preglobalization phase was marked by rapid expansion of the public system of higher education, promotion of social and occupational mobility and equality, the postliberalization phase is marked by the expansion of the unaided and self-financing private sectors in higher education. Simultaneously, the state universities are beginning to function like the private ones as far as management of finance and its impact on the academics and students are concerned. This development is discussed later in this section. For the past few years the central government has encouraged the universities to raise funds on their own and to go for fund diversification. The state-supported quality institutions have been allowed to substantially raise the tuition fees, receive donations, and enter into linkages with the industry and the corporate sector. They can also give their land for use by the private corporate sector such as IBM. Although this is happening on a small scale, the impact of this on the faculty needs to be examined.

Example no.1 A woman engineer has a Btech from IIT and also a doctorate from IIT. Her doctoral research was completed with a grant from IBM, which had set up its India Research Laboratory on the campus of IIT, New Delhi in 1998. The scholars there work on the same cutting-edge technology as in the United States of America. This young computer engineer works in the same IBM laboratory. The working conditions and salary are far superior and higher than of the faculty members employed by IIT within the same campus. Her father is a renowned former director of IIT. In spite of her background she is not willing to consider working as a faculty member of IIT because the salary is too low in comparison to what IBM pays her.

Does it benefit the host institutions academically? There is definitely a financial or commercial benefit but the academic one is suspect because it was reported in the newspapers that the faculty of IIT, Delhi, have reportedly objected to the lack of access to the IBM facilities.

As mentioned above, higher education in the preliberalization phase in India was funded, controlled, and regulated by the state until the early1980s,[21] when the mix of popular democratic politics and the rising demand for engineering and medical education led to the establishment of some private unaided/self-financing colleges in the provinces of Maharashtra and Karnataka. From the early 1990s, the burgeoning demand for professional degrees in medical, engineering/technology, computers, and management education is being met by the establishing private unaided/self-financing colleges.[22] They offer a narrow range of subjects (Meek, 2002:15) in technical/administrative knowledge (Apple, 1997), and promotion of quality in education is not on their agenda. "The policy intention was that competition would encourage institutions to find their own particular market niche. However, it seems that institutions have been more prone to copy each others' teaching . . . profiles than to consciously diversify" (Meek, 2002:15).

According to Anandkrishnan (2004) the private technical education system in India is the largest in the world, and the growth of higher education in the past 15 years has been mainly in the private sector. They seem to fulfill the demand for undergraduate professional courses in engineering/technology, medicine including dental education and health sciences, management, computer and IT education, mass media and communication, teacher education, and so on. Most of these are in the southern and southwestern states of Karnataka, Andhra Pradesh, Tamil Nadu, and Maharashtra. Other provinces are following suit. They are quick to respond to the demand for new programs, though in a limited number of subjects. As a result, their number has increased so much so that they form a majority of the undergraduate colleges in India. For example, in 2002, of 977 undergraduate engineering colleges, 764 (78.2 percent); of 1,349 medical colleges, 1028 (76.2 percent); of 505 management institutions, 324 (64.2 percent); and of 1,521 teacher education colleges, 1,038 (67.4 percent) were private and most of them substandard.

The growth of private education has contributed to the increase in undergraduate enrollment in higher education mainly in the application-oriented science and professional subjects that are being offered in the colleges of arts and sciences. In fact, they offer what are known as the emerging application-oriented science and management courses in microbiology, biochemistry, business administration, and computers.[23] For example, in Tamil Nadu the number of self-financing colleges in arts and science (they offer applied science and arts courses) has increased from 54 in 1993–1994 to 247 in 2000–2001, while the government colleges increased only from 56 to 60 and aided colleges from 132 to 133 (Bhattacharya, 2004:218). The proportion of women also increased from 42.89 percent to 51.07 percent in the private colleges during this period.

The politicization of higher education has two phases. From 1950 to 1990, the politicians in the government and the opposition pandered to the demands of teachers, students, and the lower level administrative staff because they were the vote banks as well as the potential entrants to their parties. Therefore, over a period of time universities and colleges became the hotbed of politics. The trade unions of teachers, students, and staff were all affiliated to different political parties and obeyed their directives. Since their sole purpose was to have a stable constituency of voters, their main concern was to please the vote banks even at the cost of equality and efficiency of the system. Therefore, over a period of time, the universities and colleges became, by and large, inefficient, unaccountable, and lacking in quality. The ruling politicians made no efforts to allow them to function autonomously, except elite institutions such as the IITs, IIMs, and the Indian Institute of Science. Therefore, the latter set of institutions has international reputation.

It is interesting that politicians did not want to protect the public sector institutions from political interference because they did not have a financial stake in it. After all it was public money. Their stake was only political, and it was convenient to control the vote banks at the cost of quality, efficiency, and accountability. In addition, no increase in tuition fees was allowed to increase for nearly 50 years and security of tenure of the teachers and the staff was at the top of their agenda.

Post-1990, the political interference was also high, but over a period of time the nature of political interference has changed. Now the politicians own the colleges or preside over the trusts that run the private colleges and institutions. So their interests are strictly commercial, they want to make profits, and, therefore, quality of education, efficiency, and accountability of teachers and staff are being redefined and have become their priorities. According to experts, quite a few of them are frauds. "Most of them are on the way up because they are owned by politicians" (Indiresan, 2004:13). Security of tenure is given the go-by while tuition fees have become very high even by international standards. Therefore, there has been a complete reversal of the priorities of the politicians.

Private institutes seem to be thriving more in the backward regions. For example, Chhattisgarh, a new state created on the grounds of its social and economic backwardness, has established about 150 private universities that are either on paper or are setting campuses outside the state.[24] One of them, Rai University, has established 19 campuses outside the state within 2 years. The other, Amity University, has 11 campuses. Both run like educational corporations, and both have set up schools and colleges in different parts of the country at a very fast pace. Most of them are headed by retired academic/administrators such as the vice-chancellors, college principals, the chairman of UGC, and so on who enjoyed the benefits of a tenured post with attendant benefits and are now getting huge monetary advantages plus providing legitimacy with them.

Some exceptional private technical institutes are paying salaries with incentives to a few well-established academics[25] to attract students, yet quality and merit are replaced by commercial interests in most of them (Anandkrishnan, 2004). Since they are self-financing, they are expected to and are allowed to fix the tuition fees so that they can appoint and equate adequate number of teachers and nonacademic staff and also pay them regular salaries. However, the governments and the judiciary have had to intervene to fix the fees, which were unreasonably high. In addition, some manage to take donations or unaccounted payments from the students' family. They are very high compared to the fees levied for self-financing courses in the public universities.

Again, the criteria for regulating academic work and for assessing merit are changing. Furthermore, although they are expected to accept the recruitment criteria and the salary scales of the public sector, there is no transparency in recruitment. While the qualifications, as set down by the affiliating university, may or may not be followed in hiring teachers, the salaries are negotiable; and the teachers are not paid according to their qualifications. They are more likely to be given contractual assignments and paid fixed sums as salaries or paid by the hour. Sometimes they are paid less than what was agreed to when they signed their contracts. Qualified teachers move from one institution to another in search of better salary and secure service conditions. In general, private institutions are reluctant to employ full-time faculty. The teachers are hired on contract, whether full-time or part-time. Lets look at example 2:

Example no.2 About 4 years ago, a private institution of technology in a very posh building on the outskirts of New Delhi employed a young teacher with an MPhil. degree and gave a salary equivalent to teachers in the public sector. However, in less than a year, his service conditions were changed, salary reduced, and a fixed sum given as salary. He was also put on a contractual assignment. He quickly left the job and joined an international NGO.

The salaries are not competitive and there is lack of transparency. Salaries may not even be paid on time. In spite of levying very high fees they may or may not pay the teachers as per rules, and they are paid INR 2,000–3,000 a month as a salary on the basis of number of hours of teaching, which is also regulated.[26] The provincial governments have also approved salaries for these teachers and put a maximum limit of INR 5,000–8,000 per month.

The emphasis on accountability, performance, and efficiency is more important than effective teaching-learning. One would expect that the students' academic interests be of prime importance in these institutions and those of teachers of secondary importance. It is not necessary that the academic interests of the students are protected as is clear from the example given below.

Example no.3 A young statistics teacher, with a doctorate degree, worked in a private unaided Institute of Technology also near New Delhi. She narrated her negative experience of being an honest and sincere teacher. She preferred to start the class on time and to show latecomers the exit if they exceeded the time limit of five minutes. But the management told her that she cannot do that and that students were allowed to come as late as 15–20 minutes. And if they distract her she is not to reprimand them or ask them to leave the class. When she complained about it the management said that she was upsetting the arrangements in the college. When I asked her why she does not leave she said that there is no job in the public sector colleges and universities. She had worked in a national institute of education as a researcher where her job experience was good, but the salary was not enough to maintain even a single person in New Delhi. The present institute was giving her a regular salary of a lecturer in the public sector, and she would leave as soon as she could get a better opportunity.

Since the teachers are not allowed to unionize, one would expect high quality education due to lack of tenure and security of teachers and the workload fixed by the management. But this does not seem to happen. Although the private sector has expanded, the academic community is also unrecognizable since the teachers are put under more rigid controls, have to work with a tightly regulated academic timetable, and are given very little time for research, reflection, and thinking. The private institutions, therefore, do not offer the major attraction of the profession that is to pursue research in addition to teaching (Altbach, 2002:2). The existing binary division is further heightened in the HES (higher education system) wherein most of the private self-financing unaided institutions are only undertaking undergraduate teaching in professional subjects/disciplines such as engineering, medicine, management, communication and mass media, computer education, and so on. These are the emerging/applied areas that are directly linked to the job market. The humanities and the social sciences continue to be taught in public sector higher education. Research also remains in the public universities.

It is not that the government is withdrawing from the higher education sector but the funds allocation to the colleges and universities have either been reduced or have been allowed to stagnate, and it is not sufficient to offset the rise in costs. The state universities are increasingly becoming like the private institutions, although they raise funds using various strategies such as popular self-financing courses and distance education programs—and by renting buildings for nonacademic purposes. Distance education is one of the popular ways of making money, and the funds raised may be more than the annual budget sanctioned by the state. Even then there is no long-term vision or plan to set up a corpus fund.

The provincial governments have become very stringent in allotting funds for teacher recruitment, and they do not encourage universities to employ full-time faculty with funds they may raise themselves. There is a freeze on faculty appointments. Teachers are not being retrenched, but the positions of those who retire or leave the university are not filled and lapse after a few years. For example, a well-known university in western India has succeeded in raising funds on its own and deploying them for development of the campus and the teaching programs. However, the state government has put a cap on the recruitment of teachers and does not allow even the vacant posts to be filled. The university now hires teachers on fixed-term contracts. In another state university in a northern state, there were seven full-time permanent teachers in the Department of Anthropology. Three of them retired in the past 3–4 years, but their positions have not been filled. Out of the remaining four teachers another one plans to retire in 2005. That leaves three full-time teachers. In addition, two ad hoc teachers are hired and paid wages on an hourly basis. What they are paid is less than the remuneration of the lower level administrative staff in the same university. Two of the retired teachers have also been asked to teach for a token honorarium. The same situation is seen in other departments whether they teach "unproductive" or "productive" subjects.

For the past 10 years, most provincial governments have frozen new recruitments or have allowed temporary ad hoc appointments.[27] According to the National Institute of Educational Planning and Administration (NIEPA), there are at least 50–60 vacancies in every university. It is anticipated that several teachers will retire in the next few years. It will be difficult to replace them with young and well-qualified teachers.[28] In 1998–1999, the Andhra Pradesh government announced its intention to allow 200 posts of history teachers in higher education to lapse, that is, as and when old teachers' retired new teachers will not be recruited, so that more money could be diverted to technical and marketable education. The Indian History Congress (IHC) in its annual meeting held that very year took note of this and passed a Resolution opposing it.[29] The History and Archeology Curriculum Development Committee of the UGC took note of the tendency of the governments in the southern states to close down history departments and warned against this trend. In 2004, the UGC issued a public statement that the situation is going to be very critical if steps are not taken to recruit more teachers in the public sector higher education system (*The Hindu*, 2004). In some public universities, the specialized departments have only one teacher left. How can these departments promote teaching and research.[30] This has an impact on the quality of state universities.

Furthermore, savings are effected through hiring part-time faculty, which has an impact on the quality of teaching and research, for example, private colleges/institutions are resorting to this strategy to save expenditure on higher education. While it reduces the financial investment it also impacts adversely on the quality of teaching, research orientation, and capacity of the institution. Teachers in the universities now can be divided into those who hold full-time permanent positions, full-time on contract and part-time on contract, with the first category of teachers becoming fewer and fewer (Altbach, 2002:11).

Additionally, public universities have established new departments to introduce these programs at first- and second degree levels, especially in computers and management on self-financing basis. In New Delhi, the government has established a university that functions on self-financing basis. Gradually, in most state universities, the permanent positions are decreasing and the number of temporary/contractual positions is increasing. Thus, there are two types of teachers: (i) those who are paid regular salaries and hold tenured positions and (ii) those who hold short-term jobs with minimal pay. Thus, the academic community is a fractured community within these universities and also across universities. The salary of the teachers in the second category is not sufficient. These staff have other sources of income such as private tuition and coaching. This applies to women more than men who accept the low salaries before marriage or whose husbands are working in the same place. The *gendered implications* of this phenomenon need further exploration.

In some universities, in the new science disciplines, there are fewer core, full-time staff in comparison to the project associates and adjuncts from the private sector. There is another divide within the campuses of the public universities and institutions. As mentioned earlier, IIT Delhi has allowed IBM to set up a fully funded facility within its campus. The working conditions for research and the salaries of the faculty are far better and higher than those who are appointed in the IIT. What is the impact of this on the morale of the IIT faculty and on the future of IITs? The same faculty members who work in the IBM facility would have searched for jobs in the IITs. This has added to the quality of the IBMs and has made them the premier institutions.

Issues and Problems

Although the Indian public higher education system has not completely succumbed to market demands, the shift has an impact on the academic profession in terms of what makes good teachers and what is the place of teaching and research in higher education. The central universities are still able to retain a sense of their traditional role and values attached to higher education. They allow academic space to the faculty though they are dwindling. Yet they may not able to retain the best teachers.

Example no.4 Two faculty members both in their forties, and bright intellectuals, have recently left my university which is considered to be the premier university in India. One of them, a political scientist, has joined an international agency which works in the social sector. The other one, an international relations expert, has become the head of a renowned and old residential school for boys. Both of them have opted for better working conditions, challenging work environments, and salaries that no one can ever dream of achieving in higher education.

The impact of reduced/stagnant state support is that so far as teaching is concerned, the market is pushing the demand for specific courses and programs. This is creating a division between productive and unproductive subjects and knowledge; and between teaching and research. Knowledge/disciplines are commodified. The earlier hierarchy between humanities and social sciences, on the one hand, and pure sciences, on the other, has changed that they are further relegated to a lower position than the job and application-oriented professional skills cutting across the earlier disciplinary divide. Therefore, the "technical/administrative knowledge" (Apple, 1997) represented by the professional subjects of engineering, information technology and computer education, management, communication and mass media, medicine, etc. have become the most important. Further, evaluation of teaching programs and subjects is being used as an instrument to close down departments which are not offering market-oriented courses. The impact of these developments on women and their entry as students and teachers in higher education is yet unknown. Women have generally enrolled in the subjects which do not have a market demand and have also been employed as faculty in them (Chanana, 2004). It is also known that women tend to cluster in temporary and low paid jobs. The implications of this for women deserve serious consideration.

The result of the interaction of the university with industry, international development agencies and nonprofit organizations in developed countries has been that the faculty members, who receive huge research grants or provide regular consultancy and have market-friendly "visible" research profiles, spend less time in their departments and institutions as they are in a position to generate funds for the universities and are given time off from teaching duties. This is leading to a division of the academic staff into those who teach and those who research, those who are visible and those who are not visible. This division has started up already in India in the institutions of science and technology when industry, so far, has done little to support the growth of higher education.

The self-financing of education in the public and private institutions of higher education is meeting an unmet demand for higher education in India. The fact that students are willing to pay very high fees to acquire the qualifications and skills needed in the market indicates the need for it which the government alone could not have met. It also demonstrates that politics should have been kept out of higher education to protect quality, accountability, and efficiency. But we have not learnt any lessons to delink education from politics. Now the ownership has been given to the politicians which again impacts adversely on the quality of teachers and the teaching-learning process though for different reasons.

There is no requirement to set up trusts and endowments with dedicated funds before being allowed to start educational institutions, or to have revolving funds for giving a better deal to the teachers. The process of receiving endowment funds from the alumni started in the IITs but the government stopped it midway (Chanana, 2003). The public universities which are raising funds through distance education, etc. have not been provided any policy guidelines on how to invest them for future academic use nor do they have any long-term vision. While privatization has satisfied the demand for undergraduate professional education, it does not offer any scope for graduate education, with few exceptions, and research. It is still not the first choice for students and teachers.

While the unbridled and unregulated expansion of expensive private institutes is a matter of concern the public universities are also running along the same lines. Academic freedom and stable career are not available (Altbach, 2002:11) in both. The academic profession is placed under great pressure to perform without concomitant working conditions and returns. The change is due to "reduced government funding; substitution of tenured positions with short-term contracts and very low remuneration; the perception of scholarship and knowledge as commodities and students as customers; the ad hoc approach to changes and lack of a future vision and state policy on higher education have been demoralizing for the academic profession" (Farnham, 1999:ix).

UNESCO had anticipated that as newer institutions in the private sector committed to specific disciplines and subject areas emerge, the public universities will no longer be the sole and legitimate source of knowledge generation. Furthermore, "they are unlikely to make contributions to the nation's knowledge base and be forerunners of new ideas which are essential to meet the challenges of a continuously changing world" (UNESCO, n.d.:5). However, the private sector in India is unlikely to present a challenge. Although, there are more sources of funding compared to only government funds but in the earlier phase there is a limitation. "The changes in research funding have been paralleled by increasing commodification of teaching, research and administration performance, which 'encourages' academics to produce particular kinds of measurable activities" (UNESCO, n.d.:47). The conventional universities in the public sector have to continue to be the providers of quality research. But how can they do so when they lack financial support and a dynamic academic faculty?

"Today universities are more dependent than ever on national governments for their budgets" (Scott, 1998:110). To quote Rowland (2000:9), "I chose to work in a university because I enjoy like many of my colleagues intellectual community. Like many others, I also feel that the space left for reflection is threatened by a pressurized working life. It is clear to me that the most important feature of any environment that promotes professional development, and thus the learning of our students, is one that includes such reflexive space."

From Past Perfect to Imperfect Future?

The academic profession is under pressure from various quarters. It faces a serious challenge to its legitimacy, as well as its survival, from within and outside the country and from within and outside the higher education system. The division is reinforced at a fast pace of

- universities into private and public;
- those that teach and those that do both teaching and research;
- the resource generating "visible" and the "invisible" faculty members; and
- subjects that are in demand in the market and those that are not.

The tension between teaching and research and the increasing distancing of teaching from research is a great matter of concern. According to Scott (quoted in Rowland, 2000:16), "[P]rofessionalisation of academic knowledge has made it increasingly difficult to regard teaching and research as harmonious activities."

The security of service has also come to be equated with inefficiency and lack of accountability. What has been lost sight of is the intervening variable of politics. It is being argued that quality and permanence/security of jobs are not correlated. "Conversely, is lack of security correlated with quality?" The battery of claims in favour of private institutions is yet to be proved. In the absence of transparency in the private sector such statements are based on expectations and unfounded assumptions, not on hard data. The higher education system is truly in a state of transition! The public sector institutions, too, are not improving and, therefore, the young academics do not see a bright future in them. The private sector does not provide a viable and even a remote second alternative. Those in the private institutions are also not satisfied—one can count several instances of exploitation of teachers in these institutions. Teachers are indeed in a catch 22 situation.

The teachers are moving from one end of extreme protectionism; lack of accountability, efficiency and quality; minimal fees and reasonable salaries with security of tenure. Now the profession is moving to extreme exploitation of the teacher and the students. It is characterized by high fees; low salaries and insecurity of service of teachers; and stress on accountability, efficiency and quality in a narrow sense.

So far the Indian academic in the public higher education system has had near perfect working conditions and environment. They were near perfect because they were (i) far below international standards; (ii) perfect because the workload, number of working days were far less than the international standards and (iii) the freedom and time to pursue research were immense. To date no teachers' union has taken upon itself the responsibility of improving the quality of teaching and making teachers accountable.

They are still living in the past perfect. The government/governments, on the other hand, think that responding to the market and changing the service conditions of the teachers will be the wave of the magic wand. The more insecure and less paid the teachers are the better! Therefore, the larger issues such as the contribution of academics to knowledge production and creation, the university as the site of "idle curiosity" and of inspiring a whole generation of students are being lost sight of because it is the market which is becoming the sole/major determinant of what knowledge is. Caught between the changes in the public and private institutions, the Indian academic is being pushed towards an imperfect future. No doubt there is an urgent need to rewrite the rules and regulations of "permanent" or "tenured" positions for which the government has to have the political will to delink higher education from politics, to have a vision and a policy frame. In the interim period, the questions are: "How will the dynamic centres of higher learning, scholarship and research be maintained if the present situation continues?" "How will the academics generate and nurture knowledge under the present conditions?"

Notes

1. By public, it is meant the universities run with public funds or the state universities, and private refers to the private colleges and universities. It does not refer to the public-private dichotomy used in the feminist discourse.
2. Central means federal government. The central universities are fully funded by the central government.
3. Section 3 of the UGC Act provides that an institution of higher education, other than a university, which is doing work of very high standard in a specific area, can be declared a deemed university by the Ministry of Human Resources and Development (MHRD). These institutions then become equivalent to a university for the purpose of awarding degrees. They have to meet certain criteria in terms of experience, infrastructure, availability of funds, quality of faculty, and so on. before they can be granted the deemed university status. Until recently the norms were very stringent and, therefore, until the liberalized regime, very few institutions were given this status.
4. Until recently, there were private-aided colleges, but now there are unaided colleges affiliated to state universities. Deemed university status can also be granted to a specialized institution by the MHRD on the recommendation of UGC/AICTE. Of late, several private unaided professional colleges have been granted this status. Thus, private institutions are the private colleges affiliated to the state universities, professional colleges as deemed universities, and the private universities. So far two provinces, Himachal Pradesh and Chattisgarh, have allowed private universities.
5. Henceforth, state university will mean the universities established by the provincial governments.
6. The government of the United Kingdom had set up a committee under the chairmanship of Mr. Richard Lambert, former editor, *Financial Times*, to suggest the future course for British universities. This is being done when a suggestion has been made to separate the research universities from the teaching universities and to adjust the funding pattern accordingly (College Post 2001: 14).
7. The Indian government has taken steps to set up additional quality control mechanisms through statutory and nonstatutory organizations such as the National Council for Teacher Education (NCTE). NCTE has been set up to regulate quality in teachers' education especially in the programs of open and distance learning. The All India Council for Technical Education (AICTE) has been established to provide checks and balances in the expanding social demand for engineering, technical, and management education. The National Accreditation and Assessment Council (NAAC) has also been established to provide accreditation to all the universities and colleges.

8. Although student loans have been provided in several countries there is a basic difference in the loans that are being provided now. Earlier student loans were considered a part of the welfare measures needed to encourage students from disadvantaged homes to enter higher education. They were part of the philosophy underlying the thinking that education is a social right. The capacity of the students to repay the loans was a sufficient condition but not a necessary one. They were also administered by the governments and the universities. Now the capacity to repay is important, and also the banks and financial institutions are disbursing the loans. The repayment methods differ but their links with income and income tax are prominent.
9. These are established as autonomous organizations by the national Parliament and the state legislatures respectively with very few being set up by both the governments. All of them receive full funding from the federal or provincial governments.
10. The products of these institutions have been rated so high that most of them go abroad after completing their education. While this has now become a matter of economics and investments only, this phenomenon used to raise critical debate in public discourse as to whether the country can afford to invest so much in the education of those who give no returns. The questions that were raised were: should their education be subsidized, should they not pay for it?
11. Unofficial estimates are that there are 11 million students in higher education now (December 2004), according to a statement to the press made by the vice-chairman of the UGC.
12. This scheme was earlier known as COSIST. It has now been renamed as ASSIST.
13. The present chairman of the UGC is taking personal interest in spreading this facility to the remotest colleges following the author's discussion with him.
14. Information given by Professor R. Pillai, vice-chairman, UGC.
15. This was mentioned in the author's discussion with Professor R. Pillai, the vice-chairman of UGC.
16. In 2001–2001, 154 college teachers received 50 percent of admissible expenditure.
17. In 2001–2002, 29 faculty members were selected under this program.
18. Most colleges provide a common staff room for all the teachers. They are not given separate rooms in which they can sit and work. It is assumed that teachers will come to the college, take their classes and go away.
19. The salary scale for lecturers is INR8,000–INR12,500; for readers, INR12,000–INR18,500; and for professors, INR16,000–INR22,000. Added to this is the dearness allowance to offset the rising price index (this is done for all employees in the public sector). Therefore, the basic salary may be much higher than is indicated in the scale. Besides that a provident fund for pension after retirement is also mandatory.
20. The public image of the teachers is that they take one position in public and another in private. It was emphasized by Dr. Arun Nigvekar, UGC chairman, in a personal interview.
21. In the 1980s, that is, in the preliberalization phase the governments of Maharashtra and Karnataka established private unaided colleges of engineering and medicine. These were granted as favors to the ruling party legislators. Although these colleges were regulated by the universities to which they were affiliated they charged very high sums of money for admission. This came to be known as the "capitation fee" as opposed to tuition fees. The former were undisclosed amounts paid in cash and without a receipt being given to the parents of the students. However, these private colleges formed a very small part of the higher education system until the 1990s. Now unaided private colleges, some of which are given deemed university status, are charging huge sums as tuition fees. The annual tuition fee for an MBA course is more than the annual salary of a university professor.
22. The usage of "self-financing" has given respectability to the private coaching and teaching centers, which until recently were referred to as the teaching shops. These self-styled academies and institutes are able to generate so much money that they run full page advertisements in the English language newspapers, something which the author's university cannot afford to do.

23. This information is given by my colleague, Professor Kulkarni, Centre for the Study of Regional Development, JNU, New Delhi, India. "I think they are run as colleges of arts and sciences in order to circumvent the approval of the All India Council of Technical Education which has stringent norms for approval and establishment of colleges and programmes. They are using a different nomenclature and the universities are allowing them affiliation."
24. This information is given by Professor Rajashekharan Pillai, the vice-chairman, UGC. "Last week i.e. in February 2005 the Supreme Court of India has declared them unconstitutional since they are not recognized by the UGC. The provincial government is looking for ways to find a way out so that the future of the students is not jeopardized."
25. This information was given by Dr. Nigvekar.
26. This information was also given by Dr. Nigvekar.
27. According to a study of trends in higher education undertaken by the Association of Universities and Colleges of Canada, in the year 2001 the number of professors who were 55 years and over was 12,000. The study estimated that Canada will need to hire 20,000–45,000 new full-time professors in 2011 (*College Post*, 2001:14).
28. This information has been provided by Professor G. D. Sharma, head, Higher Education Unit, NIEPA.
29. Personal communication from Professor Deepak Kumar, my historian colleague in the Centre.
30. This was highlighted by Professor Rajshesharan Pillai, vice-chairman of UGC, in my discussions with him.

References and Works Consulted

Altbach, P. G. (2002) "Centers and Peripheries in the Academic Profession: The Special Challenges of Developing Countries." In P. G. Altbach, ed., *The Decline of the Guru: The Academic Profession in Developing and Middle-Income Counties*. Massachusetts: Center for International Higher Education, Lynch School of Education, Boston College, 1–22.

Anandkrishnan, M. (2004) "Private Investments in Technical Education." In K. B. Powar and K. L. Johar, eds., *Private Initiatives in Higher Education*. New Delhi: Sneh Prakashan and Amity Foundation for Learning, 202–225.

Apple, M. (1997) "What Post-Modernists Forget: Cultural Capital and Official Knowledge." In A. H. Halsey et al., eds., *Education, Culture, Economy, Society*. New York: Oxford University Press, 1997, 595–604.

Bhattacharya, S. (2004) "Organisation of Higher Education: Moving Towards a System of Uncorporated Universities." In K. B. Powar and K. L. Johar, eds., *Private Initiatives in Higher Education*. New Delhi: Sneh Prakashan and Amity Foundation for Learning, 202–225.

Bhushan, S. (2004) "Improving Quality and Infrastructure to Make Indian Higher Education Globally Competitive." New Delhi, National Institute of Educational Planning and Administration (unpublished).

Brennan, J., J. Fedrowitz, M. Huber, and T. Shah (1999) *What Kind of University: International Perspectives on Knowledge, Participation and Governance*. The Society for Research into Higher Education Service, Buckingham: SRHE and Open University Press.

Buchanan, J. M., and N. E. Devletoglou (1970) *Academia in Anarchy: An Economic Diagnosis*. New York: Basic Books.

Campbell, I. (1996) "The Changing Face of Higher Education in the UK: A Revolution in Progress." Report published by the Institute of Psychiatry, University of London. Quoted in Farnham, D., ed. (1999) *Managing of Academic Staff in Changing University Systems: International Trends and Comparisons*. Buckingham: Society for Research in Higher Education and the Open University Press, 11.

Chanana, Karuna (1993) "Accessing Higher Education: The Dilemma of Schooling Women, Minorities and Scheduled Castes and Tribes in Contemporary India." In P. G. Altbach and S. Chitnis, eds., *Reform and Change in Higher Education in India*. New Delhi: Sage, 115–154. Also in *Higher Education*, Netherlands, 26, 1993, 69–92.

—— (2003) "Equity and Social Access: Receding Goals of Higher Education." Paper presented at the 2nd Meeting of the UNESCO Regional Scientific Committee for Asia and the Pacific, UNESCO Forum on Higher Education, Knowledge and Research, UNESCO, New Delhi, India, September 2003.

—— (2004) "Gender and Disciplinary Choices: Women in Higher Education in India." Paper presented at the UNESCO Colloquium on Research and Higher Education Policy "Knowledge, Access and Governance: Strategies for Change," Paris, December 1–3, 2004.

Chaudhuri, P. P. (2005) "Why We'll Be a Knowledge Superpower?" *The Hindustan Times*, New Delhi, February 19, 1, 13.

Chen, Xiangming (2002) "The Academic Profession in China." In P. G. Altbach, ed., *The Decline of the Guru: The Academic Profession in Developing and Middle-Income Counties*. Massachusetts: Boston College, Center for International Higher Education, Lynch School of Education, 111–140.

Dutta, S., B. Lanvin, and F. Paua, eds. (2004) "Towards an Equitable Information Society: The Global Information Technology Report (2003–2004)." World Economic Forum, New York: Oxford University Press.

Eustace, R (1987) "The English Ideal of University Governance: A Missing Rationale and Some Implications." *Studies in Higher Education*, 12(1):7–22.

Farnham, D., ed. (1999) *Managing Academic Staff in Changing University Systems: International Trends and Comparisons*. Buckingham: Society for Research into Higher Education and the Open University.

Garrett, R. (2003) *Mapping the Education Industry: Public Companies and Higher Education*. International Higher Education. Boston: Boston College Centre for International Higher Education, 9–10.

Gibbs, P. (2001) "Higher Education as a Market: A Problem or Solution?" *Studies in Higher Education* (March 2001) 26(1):87–94.

Indiresan, P. V. (2004) "IIMs vs HRD: Battle for Control." *The Hindustan Times*, New Delhi, April 18, 13.

Kelso, R., and C. Leggett (1999) "Australia: From Collegiality to Corporatism." In D. Farnham, *Managing Academic Staff in Changing University Systems: International Trends and Comparisons*. Buckingham: Society for Research into Higher Education and the Open University, 293–310.

Kogan, M. (1988) "Managerialism in Higher Education." In D. Lawson, ed., *The Education Reform Act: Choice and Control*. London: Hodder and Stoughton, 68–81.

Lauden, K., and J. Lauden (1999) *Management Information Systems: Organization and Technology in the Networked Enterprise*. Englewood Cliffs, NJ: Prentice Hall.

Meek, V. L. (2002) "On the Road to Mediocrity? Governance and Management of Australian Higher Education in the Market Place." In A. Amaral, G. A. Jones, and B. Karseth, eds., *Governing Higher Education: National Perspectives on Institutional Governance*. Amsterdam: Kluwer Academic Publishers, 235–260.

Nonaka, I., and H. Takeuchi (1995) *The Knowledge Creating Companies: How Japanese Companies Create the Dynamics of Innovation*. Oxford: Oxford University Press.

Norris, D. M. et al. (2003) "A Revolution in Knowledge Sharing." *Educause Review*, (September/October 2003):14–26.

Perkin, H. (1969) *Key Profession*. London: Routledge and Kegan Paul.

Rajagopalan, S. (2003) "Indians Second in Pro List." *The Hindustan Times*, New Delhi, March 31, 13.

Rowland, S. (2000) *The Enquiring University Teacher*. London: Society for Research into Higher Education, Open University Press, 108–129.

Scott, P. (1998) "Massification, Internationalization and Globalization." In P. Scott, ed., *The Globalization Higher Education*. London: Society for Research into Higher Education, Open University Press, 108–129.
The Hindu (2004) "Fill up Vacancies, Avert Disaster." University Grants Commission (UGC), Bangalore, March 21.
The New Scientist (2005) "Indra Special: The Next Knowledge Superpower." February 19, 2005.
Thakur, D. S., and K. S. Thakur (2005) "Knowledge Acquisition, Management and Sharing in Higher Education." New Delhi, National Institute of Educational Planning and Administration (NIEPA) (unpublished).
Thompson, K. W. et al., eds. (1977) *Higher Education and Social Change*. New York: Praeger.
Tilak, J. B. G. (1999) "Emerging Trends and Evolving Public Policies in India." In P. Altbach, ed., *Private Prometheus: Private higher education and development in the 21st Century*. Connecticut: Greenwood Press, 113–134.
Trow, M. (1996) "Comparative Reflections on Diversity in British Higher Education." *Higher Education Digest*, 16–23.
—— (1997) "More Trouble than It's Worth." *The Times Higher Education Supplement*, December 2, 12.
UGC (University Grants Commission) (2002) "Annual Report 2001–2002." New Delhi: UGC.
UNESCO (n.d.) "University Research Management in the Asia Pacific Region, Asia Pacific Higher Education Research Network (APHERN)." Discussion paper. Higher Education Research Management Project, Paris, UNESCO.
Veblen, Th. (1993) *The Higher Learning in America*. New Brunswick, NJ: Transaction Publishers; first published in 1918.
World Bank (2002) "Constructing Knowledge Societies: New Challenges for tertiary Education." A World Bank Report, vol. 1, The World Bank Group, January 1, 2002.

Chapter Seven
Development and Impact of State Policies On Higher Education Research in Indonesia

Jajah Koswara and Muhammad Kamil Tadjudin

Introduction

On the threshold of the third millennium, public policy decisions regarding higher education should and must respond to a wide variety of far-reaching changes now taking place throughout Indonesian society. It should be noted that the following significant shifts in society will have serious impacts on higher education: (i) social stratification; (ii) enrollment demand; (iii) cost containment; (iv) consensus on financial support; (v) concerns about quality; and (vi) technological advancement. Consequently, the national strategy objective in higher education systems is to improve the competitiveness of Indonesia's higher education system by developing institution credibility through the restructuring of the nationwide system, as well as the internal university system.

Research is one of the triad role activities (education, research, and community service) in Indonesian universities. Research activities at higher education level are, in general, established by the Ministry of National Education (MONE), Indonesia; and by the state Ministry of Science and Technology (MOST) for research activities in other departments outside the MONE. The Directorate of Research and Community Service Development (DCRSD) and Directorate General of Higher Education (DGHE) of MONE are responsible for the management of research coordination for MONE.

From the outset, before the system was firmly established, the research management contributions were significant. Apart from MONE funding, monetary contributions are also obtained from MOST to support research in higher education institutions.

The aims of research, within higher education, are focused on institutional capacity building, supporting science and technological development, and estimating how these research results may improve society as a whole.

Higher Education Development Policy

In order to address issues on research development, some history affecting education development policy, in general, are worth mentioning. In a recent publication on "Indonesian Higher Education in Time and Phenomenal" (Wirakusumah et al., 2003) three stages were indicated for higher education development: (i) random growth (before 1975); (ii) systemic growth (1975–1995); and (iii) long-term development strategy (after 1995).

Random Growth (Before 1975)

In the period of early independence (1945–1950) survival was the main preoccupation of the people. In spite of major difficulties during the War of Independence, several higher education institutions still functioned. In 1961, the Government of Indonesia (GOI) Law stipulated that "higher education universities should be established; at least one in each province." Several universities were then founded and developed without proper planning, many "sprung out" of private colleges, and many others started up with minimum staff and physical facilities. Consequently, it was understandable that up to the mid-1970s "development in higher education" was considered as "random" growth.

Systemic Growth (1975–1995)

During the period 1975–1985 the development of the higher education sector changed from "random growth" to "systemic growth," with improvement of productivity, capacity, and system development as major priorities. As for the period 1985–1995, efforts were focused on consolidating the previous achievements: (i) improvement of institutional capacity; (ii) infrastructures; (iii) management; (iv) productivity; and (v) quality. During this time considerable attention was given to the development of polytechnic education (producing highly skilled middle-level technicians) and graduate education (producing researchers and university lecturers). The higher education sector vastly expanded (from 400 to 1,300 institutions) during the 1975–1995 period mostly through an additional number of private institutions. Student enrollment increased from 1.5 million (9 percent participation rate) in 1985 to 2.25 million in 1995 (11 percent). Student enrollment was predominantly in social sciences, humanities, and education (73 percent). Only 15 percent enrolled in engineering and the other 12 percent enrolled in other "hard" sciences such as health, agriculture, and basic sciences. Sad to say, this situation is not compatible with the needs of Indonesia.

Long-Term Development Strategy (After 1995)

In the mid-1990s it was recognized that there was an urgent need to further strengthen higher education to enable the country sustain its rate of economic growth and competitiveness. In the long-term strategy, four main features were implemented:

1. *Implementing a new paradigm.* This meant restructuring the management of higher education through improved autonomy, accountability, accreditation, and evaluation in order to achieve sustainability and better quality.

2. *Improving the quality and relevance of higher education.* Improvement in higher education makes it compatible with the demand of economic and social development. The following programs anticipated this: (i) more balanced field of studies; (ii) better qualification of the academic staff; (iii) improving quality and relevance of selected fields of more competitive advantage; (iv) development of international graduate programs; (v) improving research capabilities; (vi) developing international research collaborations; (vii) dissemination of research results to small industries and business; and (viii) implementation of university autonomy.
3. *Enhancing access and equity.* This was geared toward significant increases in the gross enrollment rate (GER). The programs were as follows: (i) increase in the role of private sectors; (ii) development of polytechnics; (iii) increase in the coverage of fellowships; and (iv) development of Centres of Excellence (CEs) in a more "equitable" way.
4. *Legal status of an autonomous university.* This provided public universities with the necessary academic and management autonomy to develop themselves. The "legal status" of selected state universities was changed into an "autonomous legal entity."

During the period 1996–2001, a number of higher education institutions, especially in the private sectors expanded quite dramatically from 1,300 in 1996 to around 2,000 in 2001. In public institutions, such proliferation was not quite apparent except that in 1999 an additional 25 polytechnics, which were part of public university earlier, became separate institutions. In 1999 ten education institutes were given a wider mandate to develop knowledge-based disciplines apart from education and converted into universities. The same year the four most established universities, namely, Universitas Indonesia (UI), Institut Pertanian Bogor (IPB), Institut Teknologi Bandung (ITB), and Universitas Gadjah Mada (UGM) were designated as the first batch of public universities becoming "autonomous legal entities" (DGHE, 2003a).

A Vision for 2010: Indonesian Higher Education

In a globalized world, a nation's competitiveness is defined by its country's economic relationship with world markets, while its "products tend to come less from abundant natural resources and cheap labour than from technical innovations and the creative use of knowledge, or a combination of both" (Porter, 2003). The ability to produce, select, adapt, commercialize, and use knowledge becomes critical for sustained economic growth and improved living standards. Solow (2001) and other scholars have demonstrated the striking difference in gross domestic product (GDP) of countries that can be accounted for by their "investment in knowledge." Moreover, a nation's competitiveness can only be achieved when its citizens are well educated and are able to lead meaningful lives. A national higher education system should obviously provide students with good scientific knowledge. It should also contribute to the process of shaping a democratic, civilized, humane, inclusive society, maintaining a role as a moral force and as the bearer of the public conscience. At the end of the day higher education should educate students to lead meaningful lives.

From this perspective, the "Indonesian higher education vision, 2010" contains the following features:

- *quality* education that reflects students' needs, develops students' intellectual capacity to become responsible citizens contributing to the nation's competitiveness;
- *access* and equity, providing opportunities for all citizens to develop their "highest potential levels" throughout life; and
- *autonomy* for the tertiary education institutions coupled with accountability and supported by a legal, finance, and management structure, that encourages innovation, efficiency, and excellence. Autonomy also brings a shift in the regulatory environment, which now should encourage innovations at the level of individual institutions. (DGHE, 2003b)

Current Issues in Indonesian Higher Education

The most pressing contemporary issues in higher education in Indonesia include the following:

1. *Enrollment capacity.* At present the 98 state tertiary institutions have the possibility of enrolling only about 100,000 new undergraduate students each year and 3,000 graduate students; with the private universities enrolling approximately 250,000 students. The total number of students enrolled in state tertiary institutions is approximately 1,000,000, while there are about 2,500,000 in the private universities, bringing the total number of students in tertiary institutions to approximately 3,500,000 or a gross enrollment rate (GER) of about 14.6 percent in 2004 (DGHE, 2004).
2. *Equity and participation rate.* The economic downturn at the end of the last millennium was a "challenge" to the efforts for amplifying the rate of participation whilst taking into account equity (gender, social, and regional) in enrollment. The number of students on scholarships of some kind is only around 11 percent of the total number enrolled (DGHE, 2004).
3. *Quality of education.* The quality of education is, unfortunately, not uniform throughout the system. Usually the state universities have better quality than the private ones. An external quality assurance system in the form of a National Accreditation Board for Higher Education (BAN-PT) has been established. A Programme Review System is used to review periodically the 11,000 study programs now registered. At present about 80 percent of all tertiary study programs have been reviewed. The issue of "quality" is, of course, also related to "funding."
4. *Funding.* Sources of funding for state tertiary institutions are government budget allocations (60 percent) and tuition fees (40 percent). However, monetary contributions from other sources are very limited. The average funding per year for state tertiary institutions is only about US$1,000 per student, while the real requirement would be approximately US$2,500 per student per year.

Tuition fees for regular students in state tertiary institutions range from US$50 to US$500 per year. Many state tertiary institutions have established special extension programs with higher tuition fees in order to increase their income. Because of this shortfall, maintenance in many state tertiary institutions suffers a great deal. Most private tertiary institutions do not get government support, so that their income is almost exclusively from tuition fees, which ranges from US$500 to US$7,000 per pupil per year. With the new policies and shifting role of the Directorate General of Higher Education (DGHE), schemes of financial incentives have been introduced that are open to state and private universities and should steer institutions toward (i) quality; (ii) efficiency; and (iii) equity. These schemes are based on competitive funding among equal institutions or on a tiered competition basis.

5. *Internal administrative efficiency of the educational institutions.* The internal efficiency—especially in the private universities—is still very poor, causing a shortage of manpower in certain disciplines.
6. *Relevance of the curriculum to the needs of society.* Many university graduates are out of work and cannot find employment. The blame for this situation is put on the curriculum as being not relevant to the needs of society. At present only about 25 percent of students study the areas of engineering and science.
7. *External efficiency.* Many university graduates work in areas outside their area of educational competence. Although some feel that this shows they have been well educated to be able to work in other fields; others feel that this is a complete waste of resources, especially as there are so many engineers working outside their fields of competence. Of course, a major factor in this matter is the state of development and economic situation of Indonesia.
8. *Governance.* University governance structures at present do not have sufficient autonomy to ensure institutional integrity and to fulfill the responsibilities of policy development and resource development. Public universities are treated as part of the government bureaucracy, and private universities as part of the foundations to which they belong. New laws and regulations *should and must be enacted* to clearly define the role of leaderships in universities.

Research and Research Capability Development Programs in Higher Education

Problems in Research at Higher Education

In the 1990s, the following issues were identified as the main problems opposing university research development:

1. First is disparity of research capacity across universities and across field of studies.
2. *Low research culture among academics.* Most universities are more oriented toward teaching and learning activities, while research is more driven by the need to accumulate credit points for career development rather than scientific and knowledge advancement or learning process quality improvement.

3. *Limited research funding.* Allocated funding for research either from government budget or from donor agencies is very small compared to the need for research development.
4. Low research quality due to low research capacity and deficiencies in mastering research methodology is another issue.
5. *Lack of a research "umbrella."* Most researchers conduct their research individually without an institutional framework in funding, higher research development, or topics.
6. The number of published works in high quality journals is low.
7. There is low appreciation on intellectual property rights (IPR), which renders to the low economic potentials of research result.
8. There is poor integration of research activities with graduate education.
9. There is poor research management.
10. The number of international research collaborations is limited.
11. Number of university-industry collaborations is limited (Koswara, 2002).

Establishment of Research Culture

An initial step in developing research capabilities of university staff was hampered by the lack of research funds and by the random growth of higher education institutions. The fight for survival, or at least for existence, was extremely difficult during the early years of independence. From then onward the thirst and need for higher education—and the very limited budget provided by government—were, no doubt, the reasons why teaching activities have priority over anything else when the available budget is being allocated.

Before 1988, Koswara (1996) reported that research activities at some of the bigger universities were funded through a small portion of students' tuition and were distributed "evenly" to staff. The Directorate General of Higher Education (DGHE) allocated minimum funds for research, and their continuation was unpredictable. The funds were distributed to beginners, student's research, or staff research, and credit points had to be accumulated in order to get promotion. Although there were some rules for submitting research proposals, the procedures had not been standardized. Established in 1988, about 5 percent (US$6.5 million) of the International Bank for Reconstruction and Development (IBRD) Loan for Higher Education Development Project (HEDP) was designed for a competitive university research grant of a 1-year type. Under this project, Indonesia's first competitive research grants scheme was initiated. The funds were used for research activities and management for the first 3 years. A system and mechanism was introduced followed by a rather tight monitoring and evaluation of research results both "on site" and by reading research reports. Incentives were given in the form of an invitation to present the findings at a national seminar. The HEDP was extended for another 3 years, and research funding was increased to US$15 million.

From then on the system and mechanism of "research proposal competitions" were established over 5 years, the Multi-Year's Research Programme Hibah Bersaing (HB) or "competitive grant" funded by the Government of Indonesia (GOI) was approved. This system and mechanism were continuously improved. The HB has far greater funding, and a higher standard of selection criteria has been implemented compared with the 1-year research type earlier. The discussions at the open forum of selected research proposals were of great value to the evaluators and everybody concerned. A University Research Council (URC) was activated for reviewing proposals during this period. Evaluation before the research projects could be extended for the following year was very lengthy and took up much of the evaluator's time. But it was worth all the trouble as a "research culture" for finding the best and the most valuable projects in an open manner was created. After a research project is completed in 3–5 years, the researchers are obliged to publish results and deliver a presentation. The best 5 researchers are assigned 5 minutes each to present their findings before officials of the Ministry of Education (MOE) and other top ranking bureaucrats—as well as other researchers. The system of involving top decision makers in the technicalities of research procedures works extremely well. Hibah Bersaing (HB) is still one of the most prestigious research awards in higher education.

For the year 1990–2000, research grants available primarily for establishing "research culture and capacity building" were as follows:

1. *Young researcher.* The main purpose is for strengthening research culture. It is open to all disciplines, with a small grant of about IDR10,000,000 (US$1,200), and the duration is 8–10 months.
2. *Gender studies.* It is similar to Young Researcher, with special emphasis on introducing the importance of gender studies.
3. *Basic sciences.* It is more similar to Young Researcher, but with emphasis on basic sciences, larger grant of US$1,800 per year, and the duration may go on for 2 years.
4. *Research in education.* Main purpose is to improve competence and professionalism of teachers in teacher-training institutes and to improve learning processes. Duration and amount of grant are similar to Young Researcher.
5. *Multi-Year research or Hibah Bersaing (HB).* The main purpose is for capacity building, innovation, support of national science and technology, and economic benefit. Amount of grant is about US$5,000 per year with 3–5 years duration.

Research Achievement

For an idea on how research activities were funded and how university staffs responded through applications to the opportunities provided see figures 7.1 and 7.2 below. The number of proposals received and number of proposals funded of the Young Researchers Programme (1-year type) is shown in figure 7.1 and of the Multi-Year Research Programme is shown in figure 7.2.

Figure 7.1 shows that the number of proposals received increased very significantly, but the number of proposals finally funded was relatively the same because the amount of funds was always the limiting factor. When the IBRD loan came to an end in 1994–1995, only the leftover funds were used. Finally, in 1997–1998 DGHE allocated the double Young Researcher funds through the national budget up to the present day. Response from institutions was identified not only from the number of proposals received, and finally funded, but also from the number of institutions taking

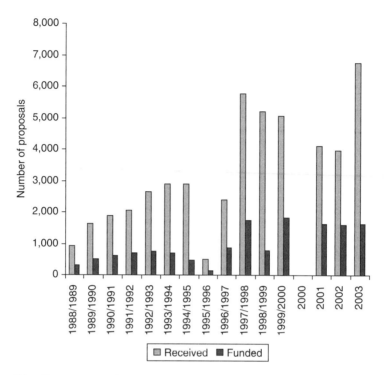

Figure 7.1 One-year research type 1988–2003
Source: Adapted from Koswara (2002).

part. In 2001, for example, 60 state and 103 private institutions were among the institutions involved in these activities, compared to less than 15 private institutions in the early 1990s. However, the number of proposals funded were 1,417 from state universities and only 219 from private universities.

As with all other research grant programs, the Multi-Year Research Programme (see figure 7.2) proposals received increased from year-to-year except when the GOI suffered financial problems, such as in year 2000—usually after 4–5 years there is a tendency for leveling off. Since this program is for more advanced researchers the following profile of around 400 research projects funded through the Multi-Year Research scheme showed as follows.

1. Each project involves 4.3 undergraduate students, 0.8 master students, and 0.3 PhD students.
2. Each project produced 1 national publication in refereed journals, 0.15 international publications, 1 national seminar, and 0.33 international seminars.
3. Each project costs IDR111.7 million or IDR42 million per year or US$5,000.
4. The profile of the principle investigators is 11 percent professors, 54 percent PhDs, 28 percent masters, and 7 percent bachelors; only 16 percent were women.

5. The profile of disciplines is engineering 20–30 percent, health and medicine 10–20 percent, agriculture 20–30 percent, basic sciences 10 percent, education 10 percent, economics, social science, law, and humanities 6 percent.
6. There is a 15–20 percent potential for obtaining patents.
7. Competitiveness is around 16 percent, terminated research (nonperformer) around 5 percent (Koswara, 2002).

The distribution of awards from 677 completed projects were as follows: 2 institutions received 110 and 129 awards, 6 institutions received between 20 and 65 awards, 7 institutions received between 9 and 17 awards, and 26 institutions received 1–8 awards. These figures clearly indicate the disparities among institutions.

Tiered Competition

Although competition is considered the best alternative in introducing the concept for getting the best out of the most available, it was found that this system could not accommodate everything. The differences in the stage of development of an institution in terms of institutional maturity, the accessibility due to geographical locations, the local importance of specific disciplines are some of the problems encountered by the diversity of Indonesian higher education institutions.

When the Multi-Year Research type was introduced, it was decided that the earlier Young Researcher scheme (1-year type) funded by a loan from the IBRD, and later

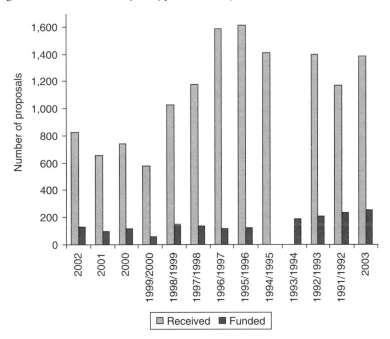

Figure 7.2 Mulit-year research type 1991–2003
Source: Adapted from Koswara (2002).

funded by GOI, was designed for only young researchers. Neither doctorate's holders are eligible, nor are staff of the four big institutions: Universitas Indonesia (UI), Institut Pertanian Bogor (IPB), Institut Teknologi Bandung (ITB), and Universitas Gadjah Mada (UGM). The young staff members of these four universities are looked after by their institutions and/or by their senior researchers.

To facilitate access and limit the influence of equalities in research competency, different fields of sciences and gender, and research capacity building, other research programs were introduced, such as basic research (since 1995), environmental studies (1994–1998), social/religious studies (since 2002), women studies (since 1995), research under the University of Research for Graduate Education (URGE) (see URGE programs, 1995–2001), Pekerti (Research/internship collaboration, since 2003), and Rapid (Research with industry, since 2004).

Other research schemes using competition procedures were also followed not only by the higher education sector but also by the Ministry of Research and Technology and all its relevant institutions with the involvement of the National Research Council (NRC). Some of the programs are Riset Unggulan Terpadu (strategic integrated research), Riset Unggulan Kemitraan (strategic integrated research with industries), Riset Unggulan Nasional (strategic national integrated research), and Riset Unggulan Terpadu Internasional (strategic international integrated research)

In the Indonesian experience, tiered competition programs in research are very useful in institution building. The program of Development of Undergraduate Education (DUE) is designed to help some universities that are already "long overdue" for being given special attention. Although it is carried out on a competitive basis, they compete among similar levels of competence. The more open competition among all undergraduate programs is quality undergraduate education (QUE). These institutions benefit from better quality through the QUE competition. Although far more proposals are qualified to be funded, some institutions are restricted in receiving support for only a certain number of study programs. The majority of the components of these programs are for "academic learning process improvement"—although a small allocation for student research is included. However, if DUE and QUE are funded through an IBRD loan, the expansion of the program such as DUE-like is fully funded by the Government of Indonesia (GOI).

University of Research for Graduate Education (URGE)

A project that directly links "research capacity" and "graduate education" is the University of Research for Graduate Education (URGE) Project. The name of the project is very appropriate since the development of "graduate" education in-country is a matter of great urgency, especially for doctorate programs or certain master programs that are still very few or not up to standard. After 2 years of very intensive preparation involving several national and international experts, the URGE Project was implemented from 1995 to 1999 through an IBRD loan and from GOI funds. The URGE Project implemented the "new paradigm," where continuous quality improvement is designed with the principle of autonomy and accountability of the university. The goal of the project is to improve the quality of graduate education in Indonesia through several specific objectives such as

1. increase competitive funding for domestic graduate education and university research activities;
2. strengthen the procedures for selecting grant and fellowship proposals;
3. integrate university research with graduate training in universities;
4. strengthen research capacity and dissemination of research findings in universities; and
5. attract highly qualified candidates for domestic graduate education.

The programs within the framework of URGE Project is believed to be the first experience within the Indonesian higher education system whereby research—and investment of relatively large grants—are provided for in the form of block grants. These programs are as follows:

1. *Center Grant*. This is with the objective to improve physical, managerial, and human resources infrastructure for high quality university research integrated with graduate training.
2. *Team Grant Programme for Graduate Research*. This is with the objective to promote high quality research activities that integrate graduate students as part of a research team.
3. *Young Academic Programme*. This is with the objective to facilitate integration of young scholars into research programs in their own universities.
4. *The In-Country Merit Fellowship*. This is with the objective to attract highly qualified students into domestic graduate education programs without any strings attached.
5. *In-Country Pre-Graduate Training Programme*. This is with the objective to improve the quality of newly enrolled graduate students, especially staff already employed by universities.
6. *Sandwich Programme*. This is with the objective to improve the quality of graduate education through collaboration with various foreign institutions.
7. *Scientific Journal Programme*. This is with the objective to improve scientific communication and to increase international recognition through improvement of scientific refereed national journals.
8. *International Research Seminar Programme*. This is with the objective to facilitate young Indonesian researchers to have international scientific community recognition, and at the same time improve dissemination of university research findings.
9. *International Research Linkage Programme*. This is with the objective to facilitate research centers and research groups to continue collaboration with foreign institutions.
10. *Domestic Collaborative Research Grant Programme*. This is with the objective to facilitate young researchers, from a wider range of universities distribution, to have the opportunity to do internship by conducting research at more advanced laboratories in collaboration with highly qualified research groups.

To maximize the right momentum on the availability of resources, expertise, and collaboration, the last three programs (International Research Seminar, International

Research Linkage, and Domestic Collaborative Research Grant) that were not originally planned were developed during the implementation stage (1995–1999) of the URGE Project. Funds for these three programs were allocated from the Directorate of General of Higher Education (DGHE). With the addition of the three programs, the URGE Project was extended another 2 years, up to 2001.

Student Creativity Program

Although this program is relatively new, it started in 2001, and for the benefit of undergraduate students, the roles of staff were significant in terms of directing and giving wider perceptions to students in the planning of sessions and monitoring the activities. This program consists of students' creativity on research, technology application, entrepreneurship, community service, and scientific writing. The Directorate General of Higher Education (DGHE) funded about 300 projects per year (from about 1,500 proposals) of IDR5 million per project or US$600/project, through a fully open competition scheme for students of public and private institutions.

Dissemination of Research Findings

The very small number of international and national publications from Indonesian scientists was very disturbing. It is apparently rooted in the lack of scientific writing experiences during the student periods. Lack of good scientific journals may be caused by lack of good papers, lack of appropriate research programs, and so on. In 1992 a small study on "scientific journal development" at universities was conducted by the DGHE. The result later became the foundation for establishing "scientific journal development" through (i) accreditation; (ii) funds for publication; (iii) involvement of professional society; (iv) incentive system for international seminars; (v) scientific writing; and (vi) journal management training.

Another issue is the dissemination of research findings to society. Since the role of every university in Indonesia is reflected in the triad role "education, research, and public service," responsiveness of university to society is often measured through activities in public service. The following programs, funded primarily by the DGHE, have been established in the past 10 years to improve the relevance of universities' research results to public services:

1. *Application of science and technology to society.* Application of applied science and technology is included in the criteria of selection.
2. *Voucher program.* A certain amount of money is granted to the proposal, which includes application of appropriate technology to small and medium enterprises (SME).
3. *Entrepreneurship program.* This is to help graduates survive after graduation by taking an active part in the economic activities of the society. The program consists of entrepreneurship courses and internship, on-the-job training, business consultation, job placement, and entrepreneur incubator.
4. *Multi-Year voucher.* After 3 years of cooperation with SME, the project should produce goods to be exported, or at least marketed, to other islands.

5. *Business unit development.* This program aimed to market research findings of the university. Local private businesses are expected to cofund the activities. Intellectual property rights (IPR) is part of this in this scheme.
6. *Sibermas program.* This emphasizes on the synergy between public empowerment and the role of higher education institutions in regional development—especially under the umbrella of regional autonomy.

Program No. 1, "Application of science and technology," to society was started in the early 1990s followed by Program No. 2, "Voucher Program," in 1994. This voucher program was a rather fundamental change in the history of direct appropriate technology application to small-and medium-enterprises (SME). The other programs were introduced during 1997–2001. Total number of proposals increased from around 300 in 1994 to almost 3,000 in 2003 for the 6 programs, and number of acceptances was 161 and 412 respectively. Figure 7.3 shows the development of the voucher program from 1994 to 2003. Unfortunately the number of received proposals in 1994–1999 was not available.

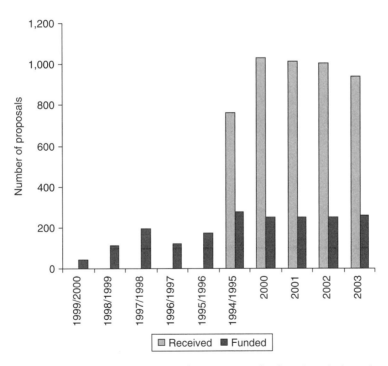

Figure 7.3 Applications of one-year type of appropriate technology (Voucher) 1994–2003
Source: Adapted from Koswara (2002).

Lessons Learned

Observing government policy on the activities of research and development schemes established by the Directorate General of Higher Education (DGHE), some lessons learned which are worth mentioning are (i) research capacity and institution building; (ii) research management and budgeting system; (iii) relevance to the society and academic community; and (iv) the impact for promoting equity and diversity.

Research Capacity and Institution Building

Government policy for faculty staff recruitment and promotions are based on civil servant regulations set up by the Ministry of Human Resources Development (MHRD). In the past recruitment was allocated from the central government based on a "quota" system for the number and field of study. The departments at the universities are often confronted with a "mismatch" between the "real need" and "staff recruited." Over the past few years, the government has implemented almost a zero growth policy, especially for the more developed institutions.

State universities propose staff career promotion, but the central government has the final say, especially for higher-level staff ranks. A reward/incentive system, which is very low, is based on government salary according to rank and does not really reflect staff performance. This system has been criticized by most of the state universities staff, without much improvement. It resulted in staff taking on extra jobs, and "moonlighting" has become a very common occurrence.

For the promotion of staff, credit points from research through publications in reputable scientific journals have become a necessity. The Directorate General of Higher Education (DGHE) refereed journals, first published in 1997, was a significant improvement as a means of giving rewards for good and productive researchers. Due to the fact that research funds, research programs, and scientific journals' accreditation system—which guaranteed well-refereed journals—were not yet established, this could not have been done before.

It would be acceptable to consider that staff with higher degrees, who would assure better knowledge, expertise, capacity, and competence, would be in line for promotion. Enrollment in the graduate programs is their main target of concern. The URGE Project with its ten programs was significant in improving the availability of graduate education in Indonesia. It is not only due to the fact that in-country programs are much cheaper than studying abroad, but also because research topics can be more relevant, and of direct benefit, to society; and the utilization of investment in research infrastructure, institution, and human resources development can be better optimized. Integration of research programs with education is another beneficial impact, not only at the graduate level but also during undergraduate studies where interaction among younger staff, students, and senior staff occur either within or between institutions. This interaction can become the basis for developing further more productive collaborations.

Research Management and Budgeting System

It has been proved that budget can be created through creation of new well-written program proposals. Since the previous programs were well planned, well implemented, and proved to be well managed by the independent reviewers/evaluators and the benefits felt by the beneficiaries, it does not matter as to whether the "chicken" or the "egg" comes first, because it involves trust. Trust of the decision makers of the systems is always at stake. Tiered competitions, regional evaluations, and regional trainings involving local expertise proved that the system worked in newly founded institutions with a big diversity of students.

Managerial capabilities should be included in program development, and decentralization made more transparent, and it should be accepted that it involves "responsibility" and "accountability." Nonetheless, although some improvement in terms of the additional number of proposals is being achieved every year available funds are still very limited. Although there is a big increase of research proposals from university staff the number of projects that can be funded has to be limited—or the unit cost has to be sacrificed—in order for the funding of more research projects.

Relevance of the Programs to Society and Academic Community

In conducting and observing all 21 programs developed by DGHE in the past 10–15 years, one can see the "silver thread" on how the government has set strategic programs that cover issues of relevance to the society and at the same time beneficial to the individuals and institutions. The following table 7.1 on relevance of current research and development programs shows the role of each program in (i) *serving the public; (ii) advancement of knowledge; (iii) education of students; and (iv) institutional and individual research capacity.*

All of the development programs are beneficial to society such as Voucher and Multi-Year Voucher programs. Small-and medium-enterprises (SME) get the benefit of technological improvement either on the production line or in marketing. Individual public opinion is that almost all the programs are beneficial to society.

Impact on Promoting Equity and Diversity

The holders of a PhD play a prominent role in research; therefore the development of in-country graduate programs is a must. On the one hand, locating financial resources for in-country graduate students enrolled in newly founded institutions is essential in facilitating the phasing out of the inequalities existing among higher education institutions. On the other hand, the recruitment of qualified staff is no easy matter. The persons proposed are sometimes not qualified to enroll in the graduate school. The program under the University of Research for Graduate Education (URGE) (pregraduate training), the program of a 1-year research type and programs such as DUE and DUE-like could be used to overcome these difficulties. It is true that in the past all government funding was targeted mainly at state universities. In the late 1990s both private and state higher education institutions were given equal treatment in the same system. The problems still exit in terms of the allocation of the government budget for education, which is very low. Therefore, combined involvement of the private sectors and the community is a must. Using regulations under autonomy/decentralization in universities and local provincial government, certain specific local

Table 7.1 Relevance of current research and development programs

No.	Research and development programme	Relevance to			Program Merit development	
		Society at large	Academic community			
		Serve the public	Advancement of knowledge	Educ. student	Inst. research capacity	Individual research capacity
01	Women studies	√			√	√
02	Application of Science and Technology to Society	√				√
03	Voucher	√			√	
04	Entrepreneurship	√		√		√
05	Multiyear voucher	√			√	
06	Business unit dev.	√			√	
07	Sibermas	√		√		
08	Student creativity	√		√	√	√
09	Young researcher		√			√
10	Basic research		√			√
11	Multi-years research	√	√	√	√	√
12	Centre grant		√	√	√	√
13	Team grant graduate res.		√	√	√	√
14	New doctor research		√			√
15	Domestic coll. res. grant		√		√	√
16	In-country merit fellowship			√		√
17	In-country pregraduate training			√		√
18	Sandwich			√	√	√
19	Scientific journal		√		√	√
20	Research publication		√		√	√
21	International seminar		√		√	√
Research relevance and merit capacity coverage (%)		45%	50%	50%	65%	90%

Source: Adapted from DGHE (2003).

needs are being encouraged. However, in terms of the diversity that exists in a country as big as Indonesia there is now a clearer picture, and it is no longer necessary that all institutions should follow the same rules. Diversity in higher education should be considered to be a contribution to the richness of society.

One criticism of the proposed schemes is that these schemes do not produce "excellent quality" science, so then why not change the funding scheme so as to fund only projects, albeit limited in number, which will produce high quality scientific results. The problem faced in Indonesia is whether the priority should be on "quality" or "equity and capacity development." The decision to choose "equity and capacity development" is a political decision necessary not only for its direct effects but also to promote national unity.

Conclusion

There are disparities in capacity, staff qualification, and research facilities among universities in Indonesia. These great differences are due, among others, to the different stages of institutional development, diversities in field of studies, or diversities in geographical locations. With equity, justice, and disparities in mind in the Indonesian higher education institutions, a tiered competition system has been developed to minimize these inequalities. The tiered competition system demonstrates significant increased individual/group participation as well as institutional participation.

The University of Research for Graduate Education (URGE) Project is an excellent project to help increase the *capacity*, and at the same time increase *participation* and *quality*, of in-country graduate programs. Government should encourage funding such activities, which is of course relatively expensive but in the long run will be a very appropriate investment, compensated by reducing expenses for "regular" overseas graduate programs.

Current research and development programs are aimed not only to develop research and human capacity, advancement of knowledge, and institution building but also to strategically serve the needs of the public. With the improvement of institutional capacity in conducting research, the need of research funds especially for newly founded institutions becomes essential. These institutions still have to develop their capabilities to a marketable level and to collaborate with other institutions on an equal basis.

The new paradigm implemented in the higher education system has certainly shown its impact on the development of some universities, especially in the private sector. Despite the positive aspects these institutions have brought to the higher education system (an excellent system for establishing methods of quality assurance) there is still a possibility of new disparities coming to light and, thus, new needs for equity treatments.

References and Works Consulted

DGHE (Directorate General of Higher Education) (2003a) "Higher Education Sector Study." Report. Jakarta, Indonesia: Directorate General of Higher Education, Ministry of National Education (MONE).

—— (2003b) "Higher Education Long-Term Strategy 2003–2010." Report. Jakarta, Indonesia: Directorate General of Higher Education, Ministry of National Education (MONE).

—— (2004) "Higher Education Policies and Programmes." Report. Jakarta, Indonesia: Directorate General of Higher Education, Ministry of National Education (MONE).

Koswara, J. (1996) "Capacity Building in University Research: The Indonesian Experience." In E. W. Thulstrup, ed., and H. D. Thulstrup, assistant ed., *Research Training for Development Proceedings of a Conference on Research Training for Countries with Limited Research Capacity*. Copenhagen: Roskilde University Press.

—— (2002) "Memorandum Akhir Jabatan 1989–2002." Direktorat Pembinaan Penelitian dan Pengabdian pada Masyarakat (Director of Research and Community Service Development Memoir 1989–2002), Dikti, Depdiknas, Jakarta, October.

Porter, M. E. (2002) "Building the Microeconomic Foundations of Prosperity: Findings from the Micro-Economic Competitive Index." In *Global Competitiveness Report 2002–2003*. Oxford: Oxford University Press.

Solow, R. M. (2001) "Applying Growth Theory across Countries." *World Bank Economic Review*, 15(2): 283–288.

Wirakusumah, S. et al. (2003) "Pendidikan Tinggi Indonesia dalam Lintasan Waktu dan Peristiwa" [Indonesian Higher Education in Time and Phenomenon]. Report. Republik Indonesia: Direktorat Pendidikan Tinggi, Departemen Pendidikan Nasional, Jakarta, Indonesia.

Chapter Eight
National Research Policy and Higher Education Reforms in Japan

Akira Arimoto

Introduction

This chapter discusses national research policy and higher education reforms in Japan. Three distinctive stages are identified—prewar, postwar, and the present—in terms of relevant developmental stages related to the chapter's themes.

As far as the national policy of science and technology is concerned, current higher education reforms are driven by the following trends: (i) substantial impact of science and technology on university reforms; (ii) globalization and the knowledge economy increasing the demands of international competition on institutions; (iii) effects of the "knowledged-based society" on university reforms; (iv) effects of the importance of knowledge development and research training on the higher education system's expectations; and (v) development of human resources. The discussion of these trends as pursued in this chapter is outlined below:

1. Social change, scientific policy, and university
2. Social changes—external pressures
3. Logic of science and scholarship in the "knowledge society"
4. Structure of science and technology policy
5. Knowledge and higher education: relationship between social condition, function, and structure of knowledge and higher education
6. Construction of the higher education system
7. Problems of the present system's focus on the graduate school
8. Concluding remarks

A total of three stages of development are distinguished in the relationship between the "national policy of science and technology" and "higher education reform," especially university reform: the (i) prewar period; (ii) postwar period; (iii) present time. Among these stages, the present time will be focused on in this chapter according to three stages of historical developments concerning Japanese university reforms from the beginning of the university establishment up until today. Only a brief outline regarding the "prewar" era follows below as many of its characteristics are also considered in relation to describing current university reform.

First, science and technology were introduced to the developing country of Japan taking the models from the developed countries in the West. The intention was to catch up with Western countries and to modernize Japan's higher education system, as swiftly as possible, by introducing advanced models of higher education.

Second, national government invited foreign scholars from the advanced countries to its newly founded universities, and in return sent their own best and brightest students to these countries in order to develop Japan's human resources, especially scientists, scholars, and researchers. This is accepted as being the first policy of human resource development at higher education level and in addition the first national policy of the internationalization of higher education in relation to the national policy of science and technology. It took some years for this policy of internationalization to come about and consequently it was changed somewhat to the self-training of scholars and students within domestic institutions.

Third, in relation to the first and second trends, it has been acknowledged that Japan tried to create its own model of higher education during the process of implanting the advanced Western models into its education system to such an extent that it was confronted with *conflicting situations* among the imported models.

This is especially true during the postwar era. It should be pointed out that a new higher education system was established during the postwar period by a great transition from the German mode prevailing during the *prewar* time to the USA mode introduced by the Ministry of Education (MOE) based on the Occupation's policy. Covering more than a 60 years' span in the postwar period, a sort of identity, or Japanese model of higher education system, was intently and constantly looked for amid the conflicting issues among foreign models themselves, particularly between the German model and the American model of higher education systems.

Through this process—over approximately 60 years—the "Americanization" of Japan's higher education system has progressed to a great extent in university reforms up until this moment in time, but not with any great success. Some of the present higher education reforms are being conducted mostly based on the USA model of higher education that was initially introduced into Japan immediately after the World War II. In a sense it is accurate to say that this "Americanization" has failed to a great degree in spite of many successive endeavors to institutionalize the model for more than half a century.

Fourth, in general, Japan was successful in catching up with the advanced model and reached the level of the Centre of Learning (COL), or Centre of Excellence (COE) by the exodus from the periphery in the sense described by Ben-David (1977). This is actually true in the fields of natural sciences and engineering, but many problems still seem to be left unsolved in the fields of humanities and social sciences in order to merit Japan's membership in the centers of excellence club worldwide. In the shadows of this gap between these two sector groups—the sector of sciences and engineering and the sector of humanities and social sciences—some conflicts are no doubt at work between the *imported cultures and the native culture*.

Fifth, the political hierarchy in the higher education system (originally made political at the outset of the country's Modern higher education system, Amano, 1993) has been constantly maintained through the postwar time even during the process of massification of higher education. In particular, the national government intentionally and consistently separated national from private higher education institutions, with an ensuing status gap between the two sectors of national and private. In addition, research universities, intended to conform to the German model from the prewar period, have had an advantageous status due to the government's constant and intensive allocation of monetary resources to these institutions, especially some of the prestigious national institutions established during the prewar time and called Key Institutions. Other institutions attempted to "catch up" with these elite institutions, or research universities, and as a result quite a few institutions have paid much attention to the "research function" of the university, whilst paying little attention to the "teaching function." The outcome was a wide gap between the institutions' research orientation and lack of teaching orientation for the masses and diversified student body (Altbach, 1996; Arimoto and Ehara, 1996).

The relationship between the national policy of science and technology and the national policy of the higher education system and its reform has been ceaselessly pursued for more than a century. But the teaching and research functions of Japanese higher education have yet to be reconciled with one another, and the tensions between the two have increased under the new pressure of emerging social changes.

Social Change, Scientific Policy, and the University

Starting with the problem of the reconstruction of the higher education system focusing on the university it is very clear that we have to start analyzing the objective situation in which the university is situated. In this analysis, it is necessary to give consideration to the situation related to the past, present, and future from a "vertical perspective" and therefore give consideration to the situation related to various differences between the Japanese education system and other education systems from a "horizontal perspective." The former is a perspective of observing the current social changes, their impacts and pressures on university, and the directions in which reforms should go. The latter is a perspective of inquiring about the problems of the Japanese higher education system compared to other systems in other countries, particularly the United States of America.

The relationship among "knowledge," "society," and "university" in the twenty-first century higher education system is changing due to the effects of social changes and the market mechanism and also the relationship between knowledge, government, society, and university.

Social Changes: External Pressure

The following three factors are estimated as being the main social changes: (1) globalization; (2) knowledge-based society; and (3) market mechanism.

Globalization

Globalization is manifestly in accordance with the "knowledge economy" and furthered by such organizations as the World Trade Organization (WTO) and the General Agreements on Trade in Services (GATS) that intend to regard education from a monetary point of view. Encouraged by this trend, international competition among higher education systems is likely to become stronger to the extent that the differences between Centres of Learning and the periphery are much more increasingly and clearly distinguished (Gumport, 2002; Altbach, 2002). Accordingly, the effects of the United States of America that is located at the head of this kind of hierarchy will be further strengthened as a main actor of "Americanization."

At the same time, as shown in the example of the Japan Accreditation Board for Engineering Education (JABEE), the reinforcement of system arrangement is clearly stressed, sooner or later, in order to meet with the global standardization and quality assurance of higher education (Onaka, 2000).

The "Knowledge Society"

It goes without saying that a system, institution, or organization that develops effectively the "knowledge function" is likely to take on further importance as the "knowledge society" itself evolves. In this context, the problems to be dealt with are development in research, the human resources development in education, and the university and society nexus in social service. We will make relevant and detailed observations concerning the development of the knowledge society in the section entitled "knowledge and Higher Education".

Market Mechanism

The term "market mechanism" means, to a great degree, the invasion of logical economics into the field of education whereby the market is strongly manipulated by a supply and demand relationship. The underlying key concept is accountability where outcome of research and education is to be satisfied so as to meet with the expectations of economic investment by the national government, local government, foundations, firms, consumers, and so on, all of whom are considered to be university sponsors. Creditability of activity related to research, teaching, and service is asked for in order to obtain external evaluation and academic personnel needs to supply sponsors with a certain message concerning the outcome of achievements. On the one hand, such factors as "accountability demands" will be given far more emphasis in the future, as responding sufficiently to the demand/expectations of sponsors is indispensable as long as a university is an enterprise based on economic investment and supported by society at large. On the other hand, the question is also asked, How far is accountability integrated in the tradition and concept of academic autonomy, which has gradually developed over as many as eight hundred years since the establishment of a prototype university?

Development of Human Resources

When reflecting on the Japanese situation concerning the relationship between the national policy of science and technology and higher education reform, additional trends of social changes must be kept in mind. The government has very strong expectations of the higher education system resulting in the development of human resources. The education system, which is committed to all life-cycle stages from "cradle to grave," implies all levels of study: precompulsory, elementary, secondary, higher, and lifelong education. The higher education level is directly committed to the function of university and therefore confronted with the massification stage of higher education development.

Referring to globalization and the knowledge economy, competition among nations for allocation of limited world resources and the development of human resources is, to a considerable degree, one and the same thing. In a knowledge society, the development and training of human resources is reliant on competent knowledge distribution in the higher education system; accordingly, teaching/learning are the basis of social development and it stands to reason that the distinguished human resources developed will be integrated satisfactorily in the "knowledge society."

The Centres of Learning (COL) with an accumulation of competitive functions of research, education, and learning are seen in the role of a "magnet"-attracting prominent human resources worldwide; such precious human resources are acknowledged, naturally enough, in such centers. As a result, a North-South problem is induced in the development and allocation of human resources to the extent that "brain drain and brain gain often take place around Centres of Learning" (Altbach, 2002; Arimoto, 2004).

These kinds of centers of learning are usually situated at the center and even right "in the realm" of higher education systems. Naturally, competitiveness and reputation of every system is internationally examined and screened more or less but henceforth every nation-state is compelled to pay much more attention to the formation mechanisms of centers of learning as well as the higher education systems.

Of course, Japan is no exception when it comes to its strong dependence on human resources development—it has a long way to go due in part to limited material resources. It would not be oversimplistic to note that "how to construct/establish a higher education system in a short-, middle- and long-term time" is keenly related to the future of a nation's development.

It may be said that this standpoint has been continuously reflected in science, technology, and higher education policies for over 130 years, since the Meiji era, when the modern higher education system was first introduced into the country.

Financial Crisis

It should be emphasized that Japan is at present confronted with a dramatic economic situation as is reflected in its debts of more than JPY700 trillion. Consequently, in relation to many fields, including education, economic rationalization is in force concerning budgets and financial contributions. In general, high quality productivity of research and education involving less budget expenditure is imperative. At the same time, social demand for university vitality as one of the substantial forces for social development is increasingly expected to take place by way of university and society nexus.

Decrease in Population

The population ratio is estimated to decrease in a long-length simulation from 1.5 million to 1.1 million for the 18-year-olds—age of university/colleges entrance. This kind of decrease in the long span is caused by birth-rate descent as has been recently shown by the 1.3 specific birth indicator. Following the basis of this trend by the end of the twenty-first century Japan's population should decrease, more or less, to 80 million people from the current 127 million (National Institute of Population and Social Security Research, 2004). Probably this trend will have great effect upon the country's future in terms of number and quality of workers and hence their education.

As far as higher education is concerned, development of human resources is thought to become a much more important problem, especially the "enhancement of quality education" to a "diminishing human resource quota."

Lifelong Learning Society

At the same time as high quality education is needed for decreasing the population, lifelong learning is also in demand by many people of the emerging knowledge society and lifelong learning society. The lifelong learning society has very much developed in accordance with such movements as UNESCO's "Life-Long Education Plan," OECD's "Recurrent Education Plan," and the Carnegie Council of Education's "Learning Society Plan." In Japan, conceptual transition was made from lifelong education to lifelong learning, and corresponding to this trend the reorganization of the Bureau, from lifelong education to lifelong learning, was made in the Ministry of Education, Culture, Sports, Science and Technology (MEXT) around 1980. This trend of stressing lifelong learning is a sort of barometer of the nation that is responsible for future education planning.

When the higher education system shifted from the massification stage to the universally accessed stage, according to Trow's theory (1974), university education was to be seen as meeting up with the perspective of lifelong learning: at this stage the university is engaged not only with traditional students but also new students, including adult students from the age of 18 to 90. In this context, transition and particularly continuity among school, university, and the society at large is questioned; and the articulation of university and school and also of university and work is brought to the fore for consideration.

Logic of Science and Scholarship in the Knowledge Society

Such a large amount of social change necessarily affects the national government's higher education policy and planning, and especially its science and technology policy. The "knowledge-based society" that is recognized to be by far the most important of these social changes necessarily has great effects on higher education reforms by means of national policies. University's response to the demand from science and scholarship means the demand derives from the emerging knowledge society, and especially the development of knowledge on which the universities' activities are substantially based and have been for many years, since their origins until today. Accordingly, this is a very natural demand, proper within academia, as a place of inquiry on the basis of the logic of science and scholarship. But if the university wants to be a place of inquiry for scholarship still more in the future, will it not self-destroy its raison d'être for not respecting the function and logic of knowledge?

When knowledge is changing in nature from Mode 1 to Mode 2 (Gibbons et. al., 1994), for example, and society is changing due to the emerging knowledge-based society, the university cannot exist by ignoring the functions of knowledge which form the basis of the university's existence in society. When knowledge itself is transformed through research, the university is expected to engage in a "reconstruction of intelligence" (University Council, 1998) with respect to teaching, service, administration, and management.

The basic trait of the knowledge-based society is that the knowledge function is changing to the extent that the transformation of the *old* knowledge society to the *new* knowledge society is taking place.

The Knowledge Function

The knowledge function, consists mainly of discovery, dissemination, application, and control, with each of them corresponding to research, teaching, service, administration, and management. The "knowledge society" or knowledge-based society means a society in which knowledge is mainly a driving force or vehicle for changing society. The university is a knowledge-based society from the day of its origin in the sense that it has a tight relationship with learning by including the function of knowledge within the university. It is easily understandable that the "university imbedded inside itself the 'stuff' of knowledge as the basis of its various activities" (Clark, 1983) and tried to organize itself to incorporate the functions of knowledge. The Middle Age University, for example, engaged in teaching as a dissemination function of knowledge, and the Modern University added research as a discovery function of knowledge and service as an application function of knowledge, with the result that the university today includes these kinds of variously accumulated functions "accompanying numerous competitions and conflicts among themselves caused by the accumulation of these functions—at the same time putting stress on research in most of all these functions" (Arimoto and Yamamoto, 2003).

Knowledge-Based Society

The university is considered to be a knowledge-based society consisting of a knowledge function, but it is different from the emerging knowledge society in the grand total of society worldwide. If we retrospectively consider the past history of higher education, separation of four developmental stages come to light: (i) Middle Age University; (ii) Modern University; (iii) University Today; and (iv) Emerging Future

University. If we consider in more detail the relationships between the characteristics of "knowledge" and "types" of university, the Middle Age University where institutionalization of science into the university had not come about and belonged to the preknowledge-based society, and the Modern University where institutionalization of science has come about so it is part of the Knowledge-Based Society 1 (KBS1), and finally the university today where science is institutionalized into not only the university but also society worldwide and belongs to the Knowledge-Based Society 2 (KBS2) (Arimoto, 2002).

These types of university such as Preknowledge, Knowledge 1, Knowledge 2, have devolved in relation to the changing characteristic of knowledge. The preknowledge society has a specific characteristic to the point that it is related to the teaching-oriented university. This type has been succeeded by the collegiate university today.

On the one hand the Knowledge-Based Society 1 (KBS1) devolved from the age of the scientific revolution—where the scientific society or scientific community was established in the university—to the age of the German model of the university, where the institutionalization of science was established within the university as an academic science (Merton, 1973). Today, the university derived from this model is thought to be the "Research University" (Geiger, 1993; Clark, 1995). On the other hand the Knowledge-Based Society 2 (KBS2) is derived from the appearance of the total knowledge-based society whereby knowledge is differentiated by Mode 1 and Mode 2 (Gibbons et al., 1994) and the distinction between both the university and the total society, both of which are related to these two modes, is borderless.

Thus far, the university's image has not been clearly established. Just like the university of today—which is made up of a mixture of the collegiate university, research university, and service university reflecting the past and traditional university models—the future university intends to create new university images out of the various existing mixtures, including the virtual university.

Structure of Science and Technology Policy

As previously discussed, national science and technology policy and relevant higher education reforms are necessary to cope with the new social changes and knowledge functions. The preceding research which was conducted with regard to the formation of "Centres of Learning" (COL) discussed the following important factors:

- Social system
- Culture and climate sustaining science and technology
- Higher education system
- Government policy of science and technology
- Traditions and characteristics of the scientific and academic community (Arimotos, 1994:216–217)

Totally speaking, the weight of the social system and the relevant conditions for promoting scientific and academic productivity are clearly recognized therein. In this context, we should pay much more attention to government policy.

Recently, the Japanese government introduced the Science and Technology Basic Law and Plan and outlined higher education reform plans such as the 21st Century

Centres of Excellence Programme, "Centres of Learning Programme," and the "Incorporation of National Universities."

Science and Technology Basic Law

A policy of the creative country with the intensive promotion of scientific and technological productivity "Kagaku Gijutsu Souzou Rikkoku" was introduced accompanying the "Kagakugijutsu Kihon-hou," or the "Basic Law of Science and Technology," which was established in 1995. Based on this law, the Science and Technology Basic Plan set up in 1996, aimed at creating a nation based on the creativity of science and technology:

> This Plan was formed under the Science and Technology Basic Law (ct 130, Nov, 15,1995) and was enacted to aim at a nation based on the creativity of science and technology; in order to encourage comprehensive and systematic policies for the promotion of science and technology, such as the promotion of scientific research activities at universities, the plan is formulated to materialize the science and technology five-year policy (from the fiscal year 1996 to 2000) with the following ten years in view. (Ministry of Education, Culture, Sports, Science and Technology (MEXT), 1996)

This plan is involved in the development of research and education in higher education institutions (H. Osaki, 1999:334–338.) as follows: (1) Stress on competition and evaluation in higher education institutions, the nation's investment is to be increased and its allocation to institutions is to be based on competition and evaluation; (2) Stress on the younger generation researchers with a focus on postdoctoral and graduate students at doctoral course.

This law and plan mentioned above clearly states the conversion to a principle of demand and supply reflecting the market mechanism from that of egalitarianism that lasted for about half a century after the postwar reform of the higher education system. Intensive investment of resources into institutions on the basis of merit and competition is thought to be rational, efficient, and useful in order to improve academic productivity in the Japanese higher education system and bring it up to an international level.

The Basic Plan consists of several chapters and sections: For example, Chapter 1 (Basic Idea) consists of the following sections: (a) Situations about science and technology (Retrospection of 20th Century/ Perspective of 21st Century); (b) Japan's national image and ideal of science and technology policy (toward the realization of a nation which can contribute to the world by creation and application of knowledge: creation of new knowledge/toward realization of the nation with international competitiveness and sustainable development: creation of vitality by knowledge/toward realization of the nation with high quality of life, health and safety); (c) Comprehensiveness and strategy of science and technology policy; (d) Construction of a new relationship between science technology and society (communication between science and technology and society/return to society of science and technology related with outcome by industry); (e) Outcome and problems of the First Stage of the Science and Technology Basic Plan; (f) Basic conception for promotion of science and technology (basic aim/extension of government investment and effective and efficient resource allocation).

A Vision for Universities in the 21st Century and Reform Measures: to be Distinctive Universities in a Competitive Environment

The report was released in 1998 by the University Council (UC) focusing on the following main issues (University Council, 1998):

> Chapter One: The society at the beginning of the 21st Century and a vision for universities: (a) Prospects for the society in the beginning of 21st Century and higher education; (b) Progress in higher education reform and current issues; (c) A vision for universities at the beginning of the 21st Century.
>
> Chapter Two: Reform measures for universities' individualization: (a) Cultivation of ability to pursue one's own ends—quality improvement of education and research; (b) More flexibility in the systems of education and research; (c) Responsible decision-making and implementation; (d) Establishment of a plural evaluation system. Individualization of universities and continuous improvement of education and research; (e) Establishment of a firm basis to advance higher education reform.

Policies to Promote Scientific Research

The Ministry of Education, Culture, Sports, Science and Technology (MEXT) has shown the direction to be taken for the future of scientific research by indicating three goals: "promotion of the world's highest levels of research," "creation of new scholarship" and "contribution to the society." Based on the Science and Technology Basic Plan approved by the Cabinet Meeting in March 2001 and the discussions in the Council for Science and Technology (CST) (cf. 2003), MEXT has been pushing ahead comprehensive science promotion measures under the basic policies that include (i) respect for the independence of researchers; (ii) evolution across a wide spectrum of disciplines, from the humanities and social sciences to natural sciences; and (iii) promotion of education and research in a more unified way. Specifically, measures taken by the MEXT include the following:

1. *Increase in grants-in-aid for scientific research*: This is to increase the amount of the grants-in-aid for scientific research with the aim of facilitating the significant development of scientific research based on liberal and open ideas.
2. *Fostering and securing of young researchers*: The aim is foster and secure young researchers through various support measures such as the Fellowship Programme implemented by the Japan Society for the Promotion of Science.
3. *Improvement of research organizations*: Research organizations should be improved including university faculties and graduate schools, research institutes attached to universities, and inter-university research institutes.
4. *Improvement of research infrastructure*: Improvement can be brought about through implementation more advanced high-speed networks and more improved and expanded databases in universities.
5. *Emphasis on the world's highest levels of research*: Emphasis is to be placed on basic research in the field of astronomical research, neutrino research, accelerator

science, space science, fusion research, informatics, global environmental research, Antarctic research, life sciences, and area studies.
6. *Promotion of partnership between industry academia and the public sector*: The aim is to develop a system that promotes joint research between universities and industry as well as commissioned research from private corporations, and operate centers for cooperative research.
7. *Promotion of international scientific cooperation and exchanges*: This is to promote researcher exchanges through the Japan Society for the Promotion of Science (JSPS).

Incorporation of National Universities

Results of the legislation of the "National University Corporation Law" issued in 2003, include (i) one of the most dramatic reforms of the university since the era of Meiji; (ii) the expectation that universities develop distinctive educational and research functions on the basis of their management autonomy and independence; (iii) government responsibility to support national universities in terms of promoting academic research and producing professionals with the highest capabilities. Several aspects of this new system are identified by the MEXT as follows: (i) incorporation respectively of each national university; (ii) introduction of management techniques based on private sector concepts; (iii) people from outside the university participating in the management of universities; (iv) improvement of the process of selection of the president; (v) selection of the non—civil servant image as the status of personnel; (vi) transparency and thorough disclosure of information and evaluation.

Knowledge and Higher Education: Relationship Among Social Condition, Function, and Structure of Knowledge and Higher Education

Today, the nation-state, as well as society at large, pays much more attention to knowledge, compared to the past century, and various kinds of sciences are engaged in research on knowledge. For example, academic disciplines such as philosophy of science, history of science, sociology of science, politics of science, economics of science, psychology of science are relevant cases. With this in mind this chapter deals with the problem of scientific policy and higher education reform and intends to focus on the relationship between knowledge and university by using some concepts and approaches developed in the sociology of science. For example, using the concepts such as (i) social conditions of knowledge; (ii) social functions of knowledge; and (iii) social structure of knowledge seems to be useful for the analysis of this given theme.

Social Condition of Knowledge

The social condition of knowledge is related to the area in which knowledge is defined by social change or social expectation: the area in which knowledge is defined by social institutions including the times, social groups, social forces, politics, economy, religion, and culture. For example, the Middle Ages University was developed from churches attached to the schools where university-level education was to be conducted. In the Modern University community, scientific knowledge was introduced in the university to the extent that the institutionalization of science was carried out in German universities, including the University of Berlin, for the first time in higher education history. In addition to the function of the teaching proper to the traditional university so as to realize an ideal such as the osmosis process and the pastoral care of students, the Modern University adopted functions of research and service.

Especially, it tried to integrate emerging research with teaching from a perspective of strengthening research. In spite of introducing the Humboltian ideal, discrepancy between ideal and reality was enlarged to the extent that the research paradigm prevailed to a great degree. As a result, a category of "Research University" was gradually developed in the twenty-first century as a new ideal image of the university (Geiger, 1986, 1993).

Today's university, which is basically an extension of the Modern University and incorporates the accumulation of various knowledge functions, is apparently faced with conflicts between functions and increasing difficulties for its coordination. In other words, various pressures caused by environmental changes from inside and outside the university are working together in the university today. Four factors are distinguished in a basic framework: (i) social change; (ii) national government; (iii) society = market; (iv) university. Among these, (i) is working in the other (ii) (iii) (iv), working directly to (iv) university. Corresponding to these changes, all of (ii) (iii) (iv) are enforced to change by themselves. At the same time, (iv) is considered to be important in terms of a triangle with (ii) and (iii).

If we use Clark's triangle model of relationship among government, society, and university (academic guild) (Clark, 1983), there are some structurally observable relationships: pressure from national government to university and reversal reaction from university to national government as seen in (ii); pressure from society = market to university, and reversely from university to society = market as seen in (iii). Of course the relationship between (ii) and (iii) is also useful.

Accordingly, the pressure of social changes is working directly toward the university and at the same time working through the pressures of national government, market, and within sciences (reconstruction of intelligence). On the basis of such pressure there exists great expectations of the knowledge function since the university is an enterprise based on "knowledge functions"; and through it the university is expected to coordinate the conflicts caused manifestly and latently from various functions. There are various kinds of coordinations including political coordination, bureaucratic coordination, professional coordination, and market coordination (Clark, 1983). Presently, as political, bureaucratic, professional, and market coordination work together in great force, professional coordination from inside of the university is required to work as a counterpower.

Social Function of Knowledge

The social function of knowledge with respect to the university includes research, teaching, and social service. These functions are thought to be important because they are continually passed on over the years from university to university. It should be noted, however, that as a result the prevailing research conflicts among functions increased so much that there is now a great problem inside academia on how to resolve these conflicts. As the old type of Knowledge-Based Systems 1 (KBS1) developed solely within the university, research came to be considered to be such an important element that functions of teaching and services having deep relationships with society were inclined to be ignored. Especially, in an age like today when KBS1 is emerging on the total society level, the nexus and integration of research and teaching is facing so much difficulty that the only possibility of the realization of nexus and integration is to be made by way of its coordination.

In terms of concrete coordination between research and teaching, there are theoretically three types: (i) integration; (ii) separation; (iii) separation and integration. Among these, type (i) has not been successful thus far since the establishment of the Modern University, even if it is ideal. On the other hand, type (ii) is a real type now in action, while type (iii) is an innovative type to be realized in the future in the sense that it is looking for integration of two functions but at the same time keeping them separated.

Social Structure of Knowledge

The social structure of knowledge refers to how knowledge functions within groups and organizational levels within the university community or enterprise. Reconsideration of knowledge means reconsideration of institution, organization, and group levels in the university in KBS2, facing the problems including reconsideration of chairperson, department, institute, faculty, and other problems including organizational norms, role expectation, role taking, and role play by teachers, internalization of role, socialization (scientific socialization, teaching socialization, service socialization, etc.). Furthermore, the control of knowledge in the enterprise means the control process of knowledge by governance, administration, and management. The two main control types are top-down and bottom-up.

Through such a serious control process, norms, aims, goals of groups, organization, institutions on the basis of knowledge are gradually formed and internalized by the members of the enterprise. They are embodied in their consciousness and behavior. Through the institutionalization of the norms by the groups and members of the enterprise in the forms of ethos, ideal, value, role, and so on occurs a series of behaviors of conformity and deviation among members including, for example, conformity to norm, nonconformity to norm, deviation from norm (Becher, 1989). If the allocation of new knowledge means the realization of the reform of norms and values, success and failure of this kind of reform are necessarily affected by the means of governance and administration. In this sense, the reconstruction of knowledge necessarily needs the reform of the enterprises and cannot lack the reconstruction of discipline-oriented groups like faculties, departments, chairs, institutions that are usually formed around knowledge. However, realization of this type of reform is far from easy owing to the traditional social structure and climate intrinsic to Knowledge-Based Systems 1 (KBS1).

Relation between Condition, Function, and Structure

Outlined above are the social conditions, functions, and structures that engage in mutual relationship/mutual interaction amongst themselves. Today's university is forced to make reforms under the influence of the structural conversion as a paradigm conversion of social change. It is also asked to reconsider its function and structure so as to construct a new university image.

Accordingly, we should construct a higher education image that integrates system, institution, organization, and groups by examining the relationship among their conditions, functions, and structures. During this process we will need to comment on university reforms, including policy, planning, practice, and evaluation, and we will also need to comment on reevaluating adequately the outcomes, for example, through metaevaluation. It would be appropriate to make a "grand" schema of the higher education system from the perspective of a comprehensive policy regarding knowledge to (i) clarify the national government higher education policy and plan through various proposals of higher education reforms inside and outside academia; (ii) practice them through functions of governance, administration, management; (iii) examine them through evaluation categories such as self-evaluation, mutual evaluation, third party evaluation in the evaluation system; and (iv) to make reconsideration of the given policy, planning, and practice by feedback process.

Construction of Higher Education Systems

At present, social change internationally emphasizes the importance of the following phenomena: knowledge-based society, globalization, and the market mechanism; and in addition to these, Japan must take account of economic recession, population decline, and lifelong learning. From a macroperspective, it is essential that the higher education system, traditionally responding to an industrial society, should shift to a system that responds to the knowledge-based society—that is, a transition from Knowledge-Based Society 1 to Knowledge-Based Society 2.

The higher education system developed in the age of KBS1 is to be changed to that responding to KBS2. In other words, the higher education system developed in the KBS1 which put much weight on research is to be shifted to that developing in the KBS2 which puts weight on teaching though stressing research as a basic rather than a separate function.

In a framework that includes the pressures from both social change and scientific change, we can comprehend a triangular relationship among national government, society (or market), and the university based on expectations, pressures, and control of knowledge. Some considerations are outlined below.

First, there are the expectations and pressures from the national government. It is not denied that some 200 advanced and developing nation-states in the world are increasingly investing in higher education. Japan has attained the world standardization in quantitative development of higher education by successfully catching up with the advanced countries in terms of importing their models (Arimoto, 1994). It is not an oversimplification to say that Japan is now looking for some new models after falling into the "modeless" stage. This situation concerns both research and teaching.

At first, it is understandable that research is succeeding very well as far as the factors related to the number of researchers and research productivity is concerned; due as can be seen to the fact that the "research university cluster" has progressed to a considerable degree in an international perspective. At the same time, there exist

Table 8.1 Selected countries' shares of published papers (percentages)

Year	Country (%)					
	Japan	USA	Germany	France	UK	Others
1992	9.3	36.9	7.9	6.1	8.8	31.0
1997	10.0	34.2	9.0	6.7	9.0	31.2
2002	10.2	32.0	9.0	6.5	8.7	33.7

Source: Institute for Scientific Information-National Science Indicators, 1981–2002.

Table 8.2 Selected countries' shares of citations (percentages)

Year	Country (%)					
	Japan	USA	Germany	France	UK	Others
1992	7.6	53.0	7.6	5.9	10.5	15.3
1997	8.3	49.3	9.6	7.0	11.2	14.7
2002	8.7	48.6	11.6	7.1	12.4	11.6

Source: Institute for Scientific Information-National Science Indicators, 1981–2002.

quite a few problems to be solved urgently. Especially, how to enhance "scientific productivity," or "academic productivity" (Merton, 1973; Shinbori, 1973), which is a traditionally important problem in the scientific community—and in the academic scientific community as well. In this context, it is noticeable that in Japan "research productivity" (including academic productivity, as well as research in general) has not attained the level of the United States of America, although it has attained internationally high standards, in a short period of time, following the establishment of the Modern University system.

As is shown in tables 8.1 and 8.2, which deal with indicators related to the number of published papers and number of citations of papers among scientists worldwide, Japan is competing with three countries—France, Germany, and the United Kingdom—and is still far behind the United States of America when we compare these indicators.

As for the shares of citations in *table 8.2*, for example, Japan's share is 8.7 percent in 2002, while USA's share is 48.6 percent, about 5 times that of Japan.

How to bridge these gaps will become one of the issues to be confronted by Japanese higher education policy and reform. Based on the first plan as discussed above, the second Science and Technology Basic Plan issued in March 2001 pointed out in its proposal "(1) training of researchers and engineers and university reform, and (2) training and security of engineers." Expectation for development of research including training of researchers and engineers is stressed in order to resolve the present difficult situation.

So as to go in this direction, the national government's recent planned line of conduct is involved in the intensive making of decisions in a series of policies such as:

- reinforcement of the graduate school;
- 21st Century Centre of Excellence (COE) Programme for Research (2002) and Centre of Learning (COL) Programme for Education (2003);

- introduction of the national university corporation (2004);
- the establishment of the national university corporation evaluation committee for the assessment of new national university corporation's achievement (2004) and the allocation of resources among them on the basis of its assessment; and
- intensive construction of COE institutions by introduction of the 21st Century COE Program as well as COL Program.

It should be noted that pressure will be increased for heading in these direction.

Furthermore, on the teaching side in the plan for science and technology, there will be relevant questions as to whether the university's teaching function is effective, or not, in terms of its processes and outcomes. For example, take the graduate rate as an indicator of human manpower training, it has already developed to an international standard and so it has been successful in terms of quantity. However, it is undeniable that severe problems have come to the surface in terms of quality. For example, if we pay attention to undergraduate education, student's diversification and decline in scholastic achievement and learning ability are observable not only at the international level but also at the domestic level to a considerable degree, and especially in the latter to a greater degree (Altbach, 1996; Arimoto and Ehara, 1996). This is a good testimony of how higher education policy is bringing about insufficient "qualitative" outcome in spite of bringing about sufficient "quantitative" outcome.

The problem is that Japan is floundering, not only at the undergraduate education level but also at the graduate education level, when comparison is made with the United States of America, which has attained the status of being the major Center of Learning in the World and is acknowledged to have first-class research and teaching functions in relation to all higher education systems worldwide (Clark, 1995), attracting through "brain drain" distinguished scientists, scholars, researchers, and students from all over the World. Unfortunately "brain drain" from Japan to the United States of America does not appear exceptional (National Science Foundation, 2002).

Problems of the Present System Focused on Graduate School

Institutionalization of Graduate School in the United States of America

The research mentioned above obtained some findings regarding promoting conditions of the research university as follows: climate of department; reward system; graduate education; communication network among researchers (Arimoto, 1994:230–231). In this context, graduate education is set as a focal point here among these factors because the centers of excellence all over the world have strong graduate education systems that should be taken into account, investigated, and studied by other counties as well as Japan.

Clark (1995) made an international comparative study of graduate schools in France, Germany, Japan, United Kingdom, and the United States of America and took note of the USA's early discovery of graduate schools compared with other countries worldwide. He pointed out the characteristics of these systems: German Institute University; French Academy University; English Collegiate University; American Department Graduate University; Japanese Applied University, paying attention to the connection between the department and the graduate school in the USA higher education system as a remarkable distinction among these systems.

If we make a retrospective evaluation of the USA's historical invention, it could be argued that its success rests on introducing a graduate school based on a modified German model while incorporating its strong research focus. The Johns Hopkins University, which owes a lot to President Gilman's unusual individual efforts, could successfully build the fundamental system as follows (Arimoto, 1981; Clark, 1995; Pierson, 1952): (i) maintaining the department system instead of introducing the chair system and apprenticeship; (ii) institutionalization of two tiers system with undergraduate and graduate course; (iii) introducing schooling in the degree system; (iv) seeking nexus of research and teaching; (v) introducing decentralization and competitiveness or diversification of the higher education system; and (vi) controlling inbreeding in the organization of faculty members.

As a result, the United States of America became the Center of Learning and superseded Germany. In fact, various kinds of surveys testified that "the USA overtook France, Germany and the UK, by 19th Century and early 20th Century, and finally stood on the summit of social stratification" (Ben-David, 1977).

Institutionalization of Graduate School in the Japanese University

The introduction of the German university model into the Japanese university was not well carried out because of the lack of a mature climate and progressive atmosphere for accepting it, though the same kind of intention was carried out at Kyoto University (Ushiogi, 1984). Of course, some of the factors related to the German model were introduced into Japanese universities, such as (i) chair system; (ii) apprenticeship; (iii) single tier system; (iv) degree system; (v) research orientation; (vi) bureaucracy. However, some of the important factors were intentionally neglected and not institutionalized into Japanese universities (i) competition among institutions; (ii) "privat dozent" system; (iii) "habilitation" system; (iv) control of life-long employment system; (v) academic freedom in teaching and learning; (vi) state university.

Japan conducted the graduate school reforms at the time of the postwar school system reforms, in introducing the American model. This reform was fairly effective with regard to promoting academic productivity among faculty members especially in the field of natural sciences. The faculty of engineering, which is thought to be the core part of the applied university as pointed out by Clark (1995), has increasingly developed among various other kinds of fields. Japan has shifted in terms of academic productivity from the "peripheral place" as pointed out by Ben-David (1977) to one of a center of learning as shown in the recent statistics. However, much improvement has to be made on the present situation when comparison is made of the counterpart's situation in the United States of America.

The difference between the two countries is likely to be a result of the following factors related to the insufficient introduction of the American model: (i) insufficient shift from the chair system to the department system; (ii) maintaining apprenticeship in training researchers (especially in the fields of humanities and social sciences); (iii) insufficient institutionalization of the graduate school system especially in the fields of humanities and social sciences as shown in the low graduate number of doctoral degrees; (iv) insufficient practice of schooling in relation to the degree system (probably owing to deep relationship with apprenticeship); (v) insufficient separation between the undergraduate course and the graduate course (it may be said that the single tier system is still substantially working); (vi) keeping of research orientation in relation to the concept of scholarship (Boyer, 1991); (vii) failure of controlling of inbreeding in reflection of a university version of lifelong employment system and seniority system; and (viii) lack of decentralization and diversification of higher education in the national sector mostly due to the national governmental control of the higher education system so as to make the unification of institutions and organizations.

One can observe that the international trend pays much attention to the construction and development of graduate education. As far as the Japanese case study is concerned, it is said that the gap between Japan and the United States of America, in terms of graduate education's academic productivity, is still large even if Japan is trying to close that gap as quickly as it can.

As discussed above, the USA's introduction of the two-tier system preceded about as long as 1 century compared with other countries worldwide, although it was a product of accidental discovery. But it is true to say that this discovery eventually promoted academic productivity leading to making Centres of Learning into an international perspective.

There are other differences between the two countries in addition to the differences previously discussed (Arimoto, 1994, 2003; Arimoto and Yamamoto, 2003): (i) separation versus integration of organization of research and teaching; (ii) research and teaching orientation versus research orientation; (iii) department system versus chair system; (iv) organization with openness versus organization with closure; (v) contract system versus lifelong employment system; (vi) manifest classification of institutions versus latent classification of institutions; (vii) institutional control versus governmental control.

Concluding Remarks

The relationship among the characteristics of the national policy of science and technology, higher education policy, and higher education reforms is very close, and it is worth paying great attention to it. This relationship is becoming more important in the emerging "knowledge-based society" where knowledge functions are considered to be important in not only the university but also in the society at large. In the case of Japan, the Basic Plan of Science and Technology has been introduced, and it is said that on the basis of this plan the direction of higher education reforms have been basically defined.

National government's strategic intensification of science and technology is directed at the following focal points: (i) promotion of basic fields; (ii) intensification of research and development (R&D) responding to national and social issues including the field of life science; (iii) information and communication, environment, nanotechnology and material technology, energy, producing skills, social basis, frontier; and (iv) response to the field of rapid development. These kinds of policies and plans on the system level are, necessarily, forcing the government authorities to make higher education reforms at the levels of institutions, organizations, and groups via higher education policies and plans.

A series of national policies in science and technology was directed at the enhancement of scientific productivity comparable to that of the most advanced systems worldwide, especially the United States of America during the postwar period, which has become a Center of Excellence with the historical "invention" of the graduate school. In this context, the fruitful outcome of Japan's national policy with regard to science and technology mostly depends on how to succeed in getting successful output in graduate education.

The role of the graduate school, intended to integrate knowledge discovery intended and knowledge assimilation, will increase its weight at the edge of the shift from KBS1 to KBS2, because the construction of organizations responding to knowledge's function and nature is inevitable. The United States of America invented the system of organization of graduate school before the coming of the Knowledge-Based Society 2; and through the process of institutionalization of the graduate school, the United States of America successfully realized a philosophy of making a nexus of research, teaching, and learning originating in German universities. On the other hand, Japan failed to adopt this system at the first stage before World War II, and after the war Americanization was not successful despite the fact that the United States of America tried to introduce the American model of graduate school and education. It actually meant the failure of constructing an organization for the nexus of research, teaching, and learning.

It can be said that introducing such a model seems to be fairly difficult since the construction of the enterprise pursuing the characteristics including openness, diversification, and competitiveness is necessarily affected by such factors as culture, society, and history of one's own country. At the same time, however, fierce competition with the American model is inescapable in the emerging new era in which (i) globalization; (ii) KBS2; and (iii) market mechanism take place and accordingly the international competition increases to such an extent that quality assurance of graduate education is necessarily needed for the comparison among systems and institutions from a perspective of the global standard. The problem of globalization is, more or less, considered to be the problem of Americanization in the field of graduate education as well as the scientific and academic community.

Conducting various kinds of reforms on the basis of making an analysis of the weak points vis-à-vis Japanese graduate education is probably the main issue to be dealt with next in order for Japan to form its own center of learning. It is intended, and desired, by everyone concerned to create identity, originality, and creativity proper to Japanese graduate education instead of making overconformity and assimilating the American model in the process of globalization.

References and Works Consulted

Altbach, P. G. (1996) *The International Academic Profession: Portraits of Fourteen Countries*. Princeton: Carnegie Foundation for the Advancement of Teaching.
—— (2002) "The Decline of the Guru: The Academic Profession in Developing and Middle-Income Countries." Centre for International Higher Education, Lynch School of Education, Boston College, Chestnut Hill, MA.

Amano, I. (1993) *Kyusei Semon Gakko* [Old professional schools]. Tokyo: Tamagawa University Press.
Arimoto, A. (1981) *Daigakujin no Shakaigaku* [Sociology of homo academics]. Tokyo: Gakubunsha Publising Company.
——, ed. (1994) *Gakumon Chushinchi no Kenkyu: Sekai to Nihon nimiru Gakumonteki Seisansei to sono Joken* [Study on the centres of learning: Academic productivity and its conditions in the World and Japan]. Tokyo: Toshindo Publishing Company.
—— (1996) "Cross-National Comparative Study on the Post-Massification Stage of Higher Education." *Research in Higher Education*, 25:1–22. Hiroshima: Research Institute for Higher Education, Hiroshima University.
—— (2001) "Hoshotaikei nonakano Kenkyu Hyoka: Gakumon no Ronri no Kozoto Kinou" [Evaluation or research in reward system: With focus on scientific logic and structure]. Daigaku Hyoka Kenky (Academic Evaluation Research), Japan University Accreditation Association, 16–24.
—— (2002) "Globalization and Higher Education Reforms: The Japanese Case." In U. Enders and O. Fulton, eds., *Higher Education in a Globalizing World*. Dordrecht: Kluwer Academic Publishers.
——, ed. (2003) *Daigaku no Curriculum Kaikaku* [University curriculum reforms]. Tokyo: Tamagawa University Press.
—— (2004) "Academic Productivity and Development of Human Resources in Higher Education." *Research in Higher Education*, 34:211–234. Hiroshima: Research Institute for Higher Education, Hiroshima University.
Arimoto, A., and T. Ehara, eds. (1996) *Daigaku kyojushoku no Kokusai Hikaku* [A comparative study of academic profession]. Tokyo: Tamagawa University Press.
Arimoto, A., and S. Yamamoto, eds. (2003) *Ddaigaku Kaikaku no Genzai* [Academic reforms today]. Tokyo: Toshinado Publishing Company.
Becher, T. (1989) *Academic Tribes and Territories*. London: Open University Press.
Ben-David, J. (1977) *Centres of Learning*. New York: McGraw-Hill.
Boyer, E. L. (1991) *Scholarship Reconsidered*. Princeton: Garnegie Foundation of Advancement of Teaching.
Clark, B. R. (1983) *Higher Education System in a Cross National Perspective*. Berkeley: University of California Press.
—— (1995) *Place of Inquiry*. Berkeley: University of California Press.
COST (Council of Science and Technology) (2003) *Policy for Training and Reserve of Research Human Resources so as to Promote International Competitiveness* (in Japanese). Tokyo: Council of Science and Technology.
Geiger, R. (1986) *To Advance Knowledge; the Growth of American Universities, 1900–1940*. London: Oxford University Press.
—— (1993) *"Research and Relevant Knowledge"; American Research Universities since World War II*. London: Oxford University Press.
Gibbons, M. et al. (1994) *The New Production of Knowledge: The Dynamics of Science and Research in Contemporary Societies*. London: Sage Publications.
Gumport, P. (2002) "Universities and Knowledge: Restructuring the City of Intellect." In S. Brint, ed., *The Future of the City of Intellect: The Changing American University*. Stanford: Stanford University Press, 47–81.
Merton, R. K. (1973) *The Sociology of Science: Theoretical and Empirical Investigations*. Chicago: University of Chicago Press.
Ministry of Education, Culture, Sports, Science and Technology (MEXT) (1996) "Press Release: The Science and Technology Basic Plan." June 2, 1996.
National Institute of Population and Social Security Research (2004) "Population Projections for Japan: 2001–2050." *Journal of Population and Social Security*, 1(1):1–43.

National Science Foundation (2002) *Science and Engineering Indicators 2002*. Vol. 2, Appendix Tables. Arlington: National Science Board.

Onaka, I. (2000) "Main Points of the Japan Accreditation Board of Engineering Education (JABEE) and Improvement of teaching." *Journal of Japanese Society for Engineering Education*, 48(1):21.

Osaki, H. (1999) *Daigaku Kaikaku 1945–1999* [University reform; 1945–1999]. Tokyo: Yuhikaku Publishing Company.

Pierson, W. (1952) *Yale College; An Educational History 1871–1921*. New Haven: Yale University Press.

"Science and Technology Basic Plan" (1996) Bureau of Science and Technology Policy, Cabinet Office, Government of Japan, Tokyo, July 2, 1996.

Shinbori, M. (1973) "Academic Productivity no Kenkyu" [Study of academic productivity]. *Research in Higher Education*, 1 (in Japanese). Hiroshima: Research Institute for Higher Education, Hiroshima University.

Trow, M. (1974) "Problems in the Transition from Elite to Mass Higher Education." In *Policy for Higher Education*. Report on the Conference on Future Structures of Post-Secondary Education. Paris: OECD, 55–101.

University Council (1998) "21Seiki no Daigaku-zo to Kongo no Kaikakuhosaku nitsuite—Kyosoutekikankyo no nakade Kosei ga Kagayaku Daigaku" [A vision for universities in the 21st Century and reform measures: To be distinctive universities in a competitive environment]. Report. Tokyo: Ministry of Education, Science, Sport and Culture.

Ushiogi, M. (1984) *Kyoto Teikoku Daigaku no Chosen* [Challenge of Kyoto Imperial University]. Nagoya: Nagoya University Press.

Yamanoi, A. (2003) "A Study on the System of Fixed-Term Appointments for Faculty Members: Focusing on the Process from its Introduction to Legislation." *Higher Education Research in Japan*, 1:21–42. Hiroshima: Research Institute for Higher Education.

Chapter Nine
Acknowledging Indigenous Knowledge Systems in Higher Education in the Pacific Island Region

Konai Helu Thaman

Introduction

This chapter puts forward an argument for the inclusion of Pacific "indigenous knowledge systems" in official documentation and discourse on knowledge production and dissemination in higher education; and particularly higher educational institutions in Oceania. The reasons for this are outlined below, giving examples of recent attempts to bring this about at the regional University of the South Pacific (USP), a higher educational institution that is owned by 12 Pacific Island Countries (PICs), namely, Cook Islands, Fiji, Kiribati, Marshall Islands, Nauru, Niue, Solomon Islands, Samoa, Tokelau, Tonga, Tuvalu, and Vanuatu.

For thousands of years, the communities inhabiting the small islands of the Pacific Ocean—now referred to as Oceania—used knowledge of themselves and their environments for everyday existence (trading, working, and communicating with one another) spoke their own languages, and followed the dictates of their own cultures. However, approximately 3 centuries ago European cultures, mainly foreign, penetrated into these islands greatly influencing Pacific people's everyday lives, that is, their ways of thinking and communicating with each other. Due to this intensive transformational process, Pacific people's "knowledge systems," including the values that upheld such systems, have been *marginalized* and *silenced*, despite the fact that these knowledge systems provided the solid foundations upon which Pacific communities have been sustained, and which guided these peoples in their struggle to survive the treacherous journeys that constantly threatened to put an end to their existence. However, it is only during the latter part of the twentieth century that the Pacific Island people themselves realized that their knowledge systems needed to be given world recognition and protection as they were of utmost importance for their cultural survival and continuity.

Definitions

The author's term "indigenous people" refers to descendants of the first people of a land irrespective of whether they belong to minority or majority populations. This is in contrast to what the international community stipulates for indigenous people as those who are minorities in their own land. This being said, it should be noted that in many Pacific Island countries, and territories, indigenous people often make up majority populations.

The author uses the term "indigenous knowledge systems" (IKS) here to refer to specific systems of values, knowledge, understandings, and practices developed and accumulated over millennia, by a group of human beings in a particular region, which may be unique to that group or region. This definition is inclusive of the processes, as well as methods of knowledge creation and transmission, developed over thousands of years by indigenous peoples and emphasizes technical insights and wisdom that are the result of careful observation and experimentation with natural as well as social phenomena. The author makes a distinction between indigenous knowledge systems (IKS) and western knowledge systems (WKS). WKS are centered on scientific knowledge and normally generated by universities, governments, research centers, and industry (Kolawole, 2001:13). Finally, the term "indigenous knowledge" rather than "traditional knowledge" is preferred because the latter term sometimes has negative connotations.

Western and Indigenous "Knowledge Systems"

WKS and IKS are different on contextual, substantive, and methodological grounds. Although both systems involve rigorous observation, experimentation, and validation—while WKS often claim universality—IKS are specific to the cultures they belong to. Moreover, while many "Western scientists tend to perceive indigenous knowledge [IK] as a means of resolving development problems of developing communities, indigenous people themselves see IK as part of their overall culture and therefore important for their identities and consequently their survival as people" (Dewes, 1993).

Until recently Pacific Island people have taken their "knowledge systems" for granted. Today, however, an increasing number of them, as well as others, are realizing that so much of Pacific indigenous knowledge has been lost and more is in danger of disappearing. During the past 4 decades or so, a few people, including this author, have become advocates of the use of Pacific indigenous and local knowledge (PIKS) in order to improve the outcomes of formal education. Some have begun the long journey of reclaiming and re-presenting Pacific indigenous knowledge systems for the sake of our students, our communities, and hopefully future generations of Pacific people (see, e.g., Thaman, 1988, 1995, 2003a; Bakalevu, 2003; Smith, 1999; Mel, 2003; Teaero, 2003; Nabobo, 2003).

Several factors help explain this movement. They include (i) the fact that there are literally hundreds of vibrant indigenous cultures in the Pacific Island Region today, most of which have been around for thousands of years and have a right to be recognized and valued (Thaman, 2000a); (ii) the belief among many educators that Pacific indigenous knowledge is needed to validate and legitimize (modern) educational development activities in the Pacific Region, particularly in the eyes of indigenous peoples themselves; (iii) the need to incorporate aspects of "indigenous knowledge systems" in the curricula of Pacific schools and universities in order to help make formal education more relevant and meaningful for Pacific Island learners, many of whom continue to find formal education irrelevant and meaningless (Taufe'ulungaki and Benson, 2002); (iv) the need to have the best of Pacific Island cultures reflected in the curricula of Pacific schools (Lawton, 1974); (v) the possibility that valuing indigenous ways of knowing in Pacific Island communities would lead to mutually beneficial collaboration between indigenous and nonindigenous groups and thus help improve their treatment of one another (Delors, 1996); (vi) the importance for nonindigenous people and institutions to recognize the need for indigenous people to have ownership and control of their IKS rather than by the academy, or researchers or academics, and thus respond positively to the UNESCO Rarotonga Declaration (Teasdale and Teasdale, 1992), which called for such recognition; and (vii) the need for Pacific scholars to recognize that researching and studying PIKS in their higher learning institutions is important in itself.

Indigenous Knowledge in Higher Education

Interest in indigenous knowledge systems among higher education personnel is relatively recent. Early writing on indigenous knowledge by Western scholars was often associated with agricultural and rural development (Brokensha, Warren, and Werner, 1980). Later indigenous knowledge became important in two main areas, namely in health care—where diseases that orthodox medicine has been unable to cure were treated with "ethno-medicine"—and in housing (where local materials were often seen as more heat-resistant and durable for building houses in the tropics) and helping to conserve foreign currency as well as strengthen local industries. More recently, researchers gave an outline of certain "stages" that people tend to follow before accepting and using indigenous knowledge; they include (i) awareness; (ii) perception; (iii) motivation; (iv) evidence; and finally (v) utilization. Others, as a result of their realization of the importance of IKS for modern development, have recommended the need to document and preserve indigenous knowledge in its place of origin, as well as globally, but they warn that such documentation could benefit more powerful centers of knowledge creation and preservation and might defeat the purpose of using indigenous knowledge to help the poor, the oppressed, and the disadvantaged (Agrawal, 1995).

The importance of indigenous knowledge in some areas was associated with concerns about the underachievement of indigenous students. This has been the case with recent studies of IKS in Africa, Australia, Canada, and New Zealand where IKS was seen as useful in providing alternative conceptual frameworks that enable indigenous scholars to examine their new and acquired worldview, one that was often framed and fixed by the colonial discourse of Cartesian and Newtonian dualism (Ntuli, 2002). In the Pacific Region, IKS is increasingly seen as a new paradigm that is more congruent with many indigenous educators' as well as students' way of thinking because it emphasizes complementarity and interconnectedness rather than duality (Thaman, 1992).

Disconnectedness and duality are inherited features of Pacific education systems. They have, however, proven inadequate and/or inappropriate for dealing with the nature and profile of most Pacific indigenous communities and people: mainly because their underlying values, content, processes, and conceptual structures tend to emphasize the social and cultural aspects of other groups—usually the newly arrived—rather than those of indigenous groups. In most Pacific Island communities, formal education has proven to be "extractive" in the sense that students acquire knowledge, skills, and values that often do not allow them to easily fit back into their island communities, forcing them to seek work in places that are often situated in urban areas or, as is increasingly the case, outside of the region altogether. The situation has been accentuated by a dearth of Pacific university graduates and academics that were willing to seriously question their own disciplines or fields of work, in order to see how they could better serve their indigenous home communities and the students that they send to school.

The result has often been professional and vocational inefficiency, not to mention misuse and abuse of scarce resources. Pacific Island educators are only just beginning to question the existing educational structures and are now realizing how these have marginalized indigenous knowledge systems together with the people and communities who produced such knowledge systems (Thaman, 1988, 1992, 1995; Smith, 1999; Sanga, 2000; Taufe'ulungaki and Benson, 2002).

This marginalization may be similar to that which was evident globally in the area of intellectual property rights (IPR), where debates seemed to worsen the piracy of indigenous knowledge as well as the biological resources of indigenous people. Biopiracy, Mishana (2002) suggests, is the result of a Western style of intellectual property rights rather than an absence of such a system in poor countries that are biorich. Mshana (2002) further suggests that biopiracy is a "double theft" in that it steals creativity and innovation; at the same time the owners of indigenous knowledge are robbed of economic options in their everyday survival. Recently introduced intellectual property rights systems (PIKS) may well result in the privatizing of both the physical and intellectual resources of Pacific Island communities (PICs) and the development of monopolizing new technologies (as in the development of Noni and Kava products for example).

The strong presence of Western scientific traditions in higher education teaching and research probably prevented the early development of indigenous, context-specific knowledge systems. Although most PICs had become politically independent by the mid-1980s, the decolonization process, especially in the area of education, consistently left unchallenged the models and content of curricula as well as the structures of formal educational institutions. References to indigenous knowledge continue to be scarce, just limited to discussions about making modern development more efficient and productive, as mentioned earlier. This is unfortunate as indigenous knowledge is not only linked to better and more efficient development but also to the very identities and futures of Pacific Island people themselves (see also Crossman and Devisch, 2002).

In Australia and New Zealand debates on indigenous knowledge also focused on the need for developing an indigenous-directed partnership approach to ongoing negotiations about recognizing indigenous knowledge, the ownership of such knowledge, and developing protocols from an indigenous viewpoint. Indigenous people in these countries argue that indigenous knowledge is often categorized and determined through the perspectives of nonindigenous people who rationalize colonialism and more recently, globalization. Today, an increasing number of indigenous scholars are more openly and critically examining modern development paradigms. Previously the "objects" of colonial analysis, they are now the "subjects" who are "answering back" to the different layers of European constructions and definitions of indigenous knowledge (Fatnowna and Pickett, 2002).

Pacific Interest in Indigenous Knowledge

This author's interest in indigenous knowledge systems (IKS) dates back to the early 1980s on examination of Tongan notions of learning, knowledge, and wisdom and how these might be reflected in Tongan teachers' perceptions of their role.

Since that time, a small but enthusiastic group of Pacific Island educators have researched and written about their own cultural values and knowledge systems, emphasizing the need for development in general—and educational development in particular—to group together Pacific people and their cultures for discourses and activities associated with Pacific Island development. Much of the author's own work over the past 10 years, especially in relation to the UNESCO Chair in teacher education and culture, has focused on these concerns (see http://www.usp.ac.fj/unesco chair/).

In 2001 a group of Pacific Island scholars and researchers formed the Re-thinking Pacific Education Initiative (RPEI), after an education colloquium that was hosted jointly by the University of the South Pacific (USP), Institute of Education (IOE) and the Victoria University of Wellington, New Zealand, on the theme "Re-thinking Pacific Education." The colloquium concluded that the two main reasons why, despite heavy investment in education, Pacific Island education systems were not delivering were (i) lack of ownership by Pacific people of formal education processes; and, (ii) lack of a clearly articulated vision of the role of formal education in Pacific Island development. The main challenge was seen as the need to reconceptualize education in a way that will allow Pacific people to reclaim the ownership of the education process and allow them to articulate a Pacific Vision for Education.

This challenge has proven to be difficult to address because of an apparent vacuum that exists in the theorization of Pacific indigenous knowledge systems as well as Pacific perspectives on knowledge production, transfer and dissemination, as well as other areas such as (i) politics; (ii) governance; (iii) economics; (iv) law; (v) land; (vi) environment; and (viii) health. This seems to be leading to ever-widening gaps between formal education institutions and the majority of Pacific Island people and communities everywhere.

Furthermore, it is often extremely detrimental to Pacific people to be told by some development consultants that development rules and regulations are "universal" and "legal" while indigenous people's ideas and norms are "cultural" and "illegal."

Many would be familiar with examples of failed development initiatives in Pacific Island countries (PICs) following development models that were premised on Western, scientific, reductionist ideologies. Such failed attempts in our region have pushed many countries back to the end of the "development queue," at the head of which is the Western, capitalist model of life (Esteva, 1992). Today the promise seems to be that those who can imitate this model by undergoing reforms and restructuring will find a place at the banquet table and will be duly rewarded (Hoppers, 2002:viii).

However, choosing to center one's work of teaching and research on indigenous and local knowledge systems has had both positive and negative consequences. A positive outcome was the establishment of an UNESCO Chair in teacher education and culture at the University of the South Pacific (a position that the author currently holds). The move might have caused some to realize the importance of culture, at least in the education of teachers. Other positive outcomes were

- improved students' learning outcomes;
- better contextualized teaching and learning; and
- better motivation for indigenous students to theorize and write about their own ideas and practices.

However, there continues to be real challenges as well, often involving ideological, philosophical (methodological), and application issues and questions.

Working with indigenous knowledge systems (IKS) often makes people doubt and question the ideology of rationality, the ideological basis for liberal and scientific knowledge, introduced and propagated by European colonization and missionary influence of various Pacific societies for over 2 centuries. This ideology continues to be promoted in and by Pacific formal education systems and structures today. Pacific Island scholars, therefore, would need to acknowledge IKS in their thinking as well as in their practices, in order that they may recenter policies and practices on Pacific people, and reclaim and reemphasize the vital link between Pacific cultures and the development of Pacific people, especially in education. Pacific scholars would also need to be aware of the epistemological silencing of, and continued indifference to, indigenous cultures by many nonindigenous and indigenous people alike, and the ongoing efforts by some to discourage exchange of ideas and methods about the role of IKS in addressing educational issues and problems, because there continues to be a few Pacific scholars who are indifferent toward Pacific cultures and their indigenous knowledge systems (IKS). This has resulted in some degree of cultural blindness, evident in the practices of most of our formal national as well as regional institutions, particularly those who wish to forge ahead with the logic of Western cultures and their ideologies and epistemologies.

This is reflected in the view held by some that indigenous cultures are barriers to modern thinking and development, especially economic development. As such, IKS are often marginalized and relegated to an "inferior position," pushed to a so-called informal sector, for the interest of mostly not-so-serious scholars, nongovernmental organizations (NGOs), some students, and women. This is bound to have serious consequences for Pacific people's identity formation and human development, especially among those people who live in towns and cities both in the Pacific Region and beyond.

Some Encouraging Signs

There are, however, some encouraging signs from Western scientists and scholars from whom so many Pacific people take their cues. Western scientists have been talking about traditional ecological knowledge (TEK), a system of knowledge that builds on generations of people living in close relationship with nature. In the area of climate change, for example, there is now recognition that "*indigenous climate change assessments and observation*" are built on countless generations of knowledge, since time immemorial. Scientists also agree that TEK carries within itself systems of classifications and empirical observations about the local environment and a system of self-management that governs sustainable resource use. They also agree that the responsibility for, and the carriers of, TEK lies with the elders of indigenous communities, and because TEK accumulates and adapts knowledge in a holistic manner scientists are now urging organizations such as the "Intergovernmental Panel" on "Climate Change to validate Traditional Ecological Knowledge (TEK) as a vehicle in the assessment, research, and other scientific work on "Climate Change." They also encourage similar recognition worldwide in all environmental and resource-management work" (SnowChange.Org, 2002).

It can also be argued that IKS is a form of science as indigenous medicines, fishing and farming practices, hunting technics as well as the use of fire, require an intimate knowledge of terrestrial as well as marine ecology and biology. These require skills of observation, classification, and comparison—essential features of the scientific method. There is also the need to acknowledge the increasing flow of information from the South to the North about the role of IK in various fields such as agriculture, human and animal husbandry care, the use and management of the environment, rural development, education, and poverty alleviation.

There are also encouraging signs in the area of higher education. Despite the fact that Western processes of knowledge analysis and transmission have continued to remain unchallenged, there is now a movement to reaffirm the significance of local knowledge and wisdom due to the need to "rediscover" these as a response to students' underachievement and the impact of globalization. World attention on IK has been evident in international conferences, many sponsored by UNESCO, such as the University of Hawaii's Pacific Studies 2000 Conference, and the PROAP/ACEID Annual Conference in December 2000, in Bangkok, Thailand, which examined the "interface between indigenous and global knowledge." Emphasized in these fora was the need to preserve multiple wisdoms in the Asia Pacific Region, and mainstreaming indigenous knowledge in the work of international organizations such as UNESCO and the World Bank (WB) (see e.g., Larson, 1998; UNESCO, 2001).

Within the Asia Pacific Region, the need to revalidate IKS is being approached from an educational perspective. For example, India, Indonesia, Thailand, and Vietnam are advancing innovative new curricula and new approaches to the transfer of knowledge. The aim is to link local knowledge to global knowledge and the full recognition of indigenous knowledge. The incorporation of IK into higher education curricula is one way of recognizing the importance of it especially for indigenous students. But there remains a need to develop new methods (participatory, interdisciplinary research) to elicit and generate local knowledge, as well as innovative teaching methods that involve alternative forms of knowledge transfer, and to produce teaching materials that are adapted to local situations.

There is a growing demand in the Pacific Region for education systems to be tailored to local needs and with their own identities. The EU/NZ-funded PRIDE project, based at the University of the South Pacific (USP), is largely a response to this need. Furthermore, a Review Committee has recently called on the university to be more responsive to the needs and aspirations of Pacific communities. Clearly the pressure is on for the right of Pacific communities to have more say in determining their future development.

Many Pacific Island communities have come to view higher education as the means whereby their culturally diverse communities can achieve the prizes offered in society. The establishment of institutions such as University of the South Pacific (USP) and University of Papua New Guinea (UPNG) was seen as necessary for, and compatible with, modern global consumer economies. However, they together with staff and students within them have continued to follow models and pathways set in other, mainly metropolitan, universities. It is particularly significant today that many Pacific educators and researchers are joining forces to help reclaim and rethink Pacific education in a serious attempt to weave something better and more meaningful for themselves as well as their students (see, e.g., Thaman, 1997, 1998, 2000b, 2003a, 2003b. It is also interesting to note that many of those who are involved in this movement are women.

Rethinking Pacific Higher Education and Research

The Rethinking of Pacific Education Initiative (RPEI) continues to be concerned with ensuring that Pacific values and knowledge systems underpin modern educational development and help to decolonize Pacific education as well as Pacific minds (Taufe'ulungaki, 2003). This does not mean that colonial attitudes will die as a result; far from it. But it just means that they are no longer tolerated because they undermine our confidence in our ability to do our "own thing," to write our own stories, sing our own songs, and to reimagine ourselves.

As for the author of this chapter, teaching and researching Pacific knowledge systems has been both an education as well as an art form. "In theorizing my own education I have drawn inspiration from both Western and Pacific indigenous epistemologies. My concept of education and research, known as *Kakala*, involves three major processes, namely *toli, tui*, and *luva*."

- *Toli* is the gathering of the material need for making a *kakala* such as different types of flowers, leaves etc. This process requires knowledge of and experience in picking/gathering the appropriate materials at the right time and the right place; storing them in a cool and safe place in order to ensure freshness until they are ready to be made into a *kakala*.
- *Tui* refers to the actual making of a *kakala* and requires special knowledge and skills of different types of *kakala* depending on the occasion and/or who is to wear the *kakala*.
- *Luva*, the giving away or presentation of a *kakala* to someone else symbolizes the deep values of *ofa* (compassion) and faka'apa'apa (respect) in Tongan culture in particular and in Polynesian traditions in general. There is a complex etiquette surrounding different *kakala* with some more significant than others, based on their histories and mythologies. However, different types of *kakala* are usually needed to make a "complete" beautiful garland (Thaman, 1992).

Kakala is a useful, culturally meaningful indigenous philosophy and framework for Pacific education, research, and development. It ensures cultural inclusivity and provides for ownership of the education and development process.

Over the past 12 years or so, the author has shared *kakala* with colleagues in many parts of the Pacific Region and elsewhere. This has given the people the opportunity of remembering their origins and ways of life in the Pacific Region. Albeit, working with UNESCO the author has also been able to extend this sharing to a wider Asia Pacific Region where similar concepts exist.

Conclusion

As with most indigenous ideas, *kakala* is an integrated and holistic philosophy and framework that combines professional and creative interests. *Kakala* is not only an indigenous framework of knowledge and wisdom, but it is also about art and spirituality, aspects of life that are often missing from much of what we do in the Pacific Islands in the name of "education," "research," and "development." As many artists know, art gives people soul whether the "words" are painted, carved, sung, spoken, written, danced, filmed, or performed. Some words need a brush, others a pen, and still others body/feet movement, gesture, and intonation. Art connects Pacific people to the Pacific Ocean and to each other. However, it is their cultures and knowledge systems that will sustain and help define them, as they try to weave useful and fragrant *kakala* for their journeys into the future.

References and Work Consulted

Agrawal, A. (1995) "Indigenous and Scientific Knowledge: Some Critical Comments." *Indigenous Knowledge and Development Monitor*, 3 (December):3–33.

Bakalevu, S. (2003) "Ways of Mathematizing in Fijian society." In K. Thaman, ed., *Educational Ideas from Oceania*. Suva: Institue of Eucation (IOE), University of South Pacific (USP), 61–73.

Brokensha, W. D. M., and O. Werner, eds. (1980) *Indigenous Knowledge Systems and Development*. Lanham: University Press of America.

Crossman, P., and R. Devisch (2002) "Endogenous Knowledge: An Anthropological Perspective." In C. A. O. Hoppers, ed. *Indigenous Knowledge and the Integration of "Knowledge and the Integration of Kbnowledge Systems."* Claremont: South Africa, New Africa Books, 96–125.

Delors, J. (1996) "Learning: The Treasure Within." Report to UNESCO of the International Commission on Education for the 21st Century. Paris: UNESCO.

Dewes, W. (1993) "Introduction." In S. H. Davis and K. Ebbe, eds., *Proceedings of a Conference on Traditional Knowledge and Sustainable Development*. Washington, DC: World Bank

Esteva, G. (1992) "Development." In W. Sachs, ed., *The Development Dictionary: Guide to Knowledge and Power*. London: Zed Books.

Fatnowna, S., and H. Pickett (2002) "The Place of Indigenous Knowledge Systems in the Post-Modern Integrative Paradigm Shift." In C. A. O. Hoppers, ed., *Indigenous Knowledge and the Integration of "Knowledge Systems."* Claremont: South Africa, New Africa Books.

Hoppers, C. A. O., ed. (2002) *Indigenous Knowledge and the Integration of Knowledge Systems*. Claremont: South Africa, New Africa Books.

Kolawole, O. D. (2001) "Local Knowledge Utilization and Sustainable Development in the 21st Century." *Indigenous Knowledge and Development Monitor*, 9(3) (November).

Larson, J. (1998) "Perspectives on Indigenous Knowledge Systems in Southern Africa." World Bank Discussion Paper No. 3, Washington, DC, World Bank.

Lawton, D. (1974) *Class, Culture and the Curriculum*. London: Routledge and Kegan Paul.

Mel, M. (2003) "Shifting Cultures: Mbu—A Proposal for a Pluri-Cultural Perspective to Culture in Education in Papua New Guinea." In K. H. Thaman, ed., *Educational Ideas from Oceania*. Suva, Fiji: IOE, USP, 13–25.

Mshana, R. (2002) "Globalization and Intellectual Property Rights." In C. A. O. Hoppers, ed., *Indigenous Knowledge and the Integration of "Knowledge Systems."* Claremont: New Africa Books, South Africa, 158–172.

Nabobo, U. (2003) "Indigenous Fijian Education." In K. H. Thaman, ed., *Educational Ideas from Oceania.* Suva, Fiji: IOE/USP, 85–93.

Ntuli, P. P. (2002) "Indigenous Knowledge Systems and the African Renaissance." In C. A. O. Hoppers, ed., *Indigenous Knowledge and the Integration of "Knowledge Systems."* Claremont: New Africa Books, South Africa.

Sanga, K. (2000) "Learning from Indigenous Leadership: Module 6. Pacific Cultures in the Teacher Education Curriculum Series." Suva, Fiji: IOE/USP.

Smith, L. (1999) *Decolonizing Methodologies: Research and Indigenous Peoples.* Zed London: Zed books.

SnowChange.Org. (2002) "International Indigenous Workshop." Tampere, Finland.

Taufe'ulungaki, Ana (2003) "The Role of Research: A Pacific Perspective." In Eve Coxon and Taufe'ulungaki, Ana, eds., *Global/Local Intersections: Researching the Delivery of Aid to Pacific Education.* New Zealand: Research Unit of Pacific Education (RUPE), The University of Auckland.

Taufe'ulungaki, A. M, and C. Benson, eds. (2002) *Tree of Opportunity: Rethinking Pacific Education.* Suva, Fiji: IOE, USP.

Teaero, T. (2003) "Indigenous education in Kiribati." In K. H. Thaman, ed., *Educational Ideas from Oceania.* Suva: IOE/USP, 106–115.

Teasdale, R., and J. Teasdale, eds. (1992) *Voices in a Seashell: Education, Culture and Identity.* UNESCO/IPS, Suva, Fiji.

Thaman, K. H. (1988) "Ako & Faiako: Cultural Values, Educational Ideas and Teachers' Role Perceptions in Tonga." PhD thesis, USP, Fiji.

—— (1992) "Looking Towards the Source: A Consideration of (Cultural) Context in Teacher Education." In C. Benson and N. Taylor, eds., *Pacific Teacher Education Forward Planning Meeting: Proceedings.* Suva: Institute of Education, University of the South Pacific, 98–103.

—— (1993) "Culture and the Curriculum in the South Pacific." *Comparative Education*, 29(3): 249–260.

—— (1995) "Concepts of Learning, Knowledge and Wisdom in the Kingdom of Tonga." *Prospects: UNESCO Quarterly Review of Education*, 25(4):723–735.

—— (1997) "Kakala: A Pacific Concept of Teaching and Learning." Keynote address, Australian College of Education National Conference, Cairns.

—— (1998) "Learning to Be: A View from the Pacific Islands." Report of the UNESCO Conference on Education for the 21st Century in the Asia Pacific Region, UNESCO, Melbourne.

—— (2000a) "Interfacing Global and Indigenous Knowledge for Learning." UNESCO/ACEID International Conference Panel 2, UNESCO, Bangkok, Thailand, 1–6.

—— (2000b) "Towards a New Pedagogy: Pacific Cultures in Higher Education." In R. Teasdale and M. Rhea, eds., *Local Knowledge and Wisdom in Higher Education.* Oxford: Pergamon Press, 43–50.

——, ed. (2003a) "Educational Ideas from Oceania: Selected Readings." Institute of Education, University of the South Pacific (USP), Suva, 131.

—— (2003b) "Decolonizing Pacific Studies: Indigenous Perspectives, Knowledge and Wisdom in Higher Education." *The Contemporary Pacific*, 15(1):1–18.

UNESCO (2001) "Declaration on Cultural Diversity." Adopted by the 31st Session of the UNESCO General Conference, UNESCO, Paris, November 2, 2001.

Chapter Ten
Higher Education Research in the Philippines: Policies, Practices, and Problems

Rose Marie Salazar-Clemeña

Introduction

The advent and expansion of the "knowledge society" has brought new challenges to higher education. Situated at the heart of the "knowledge society," higher education must respond to the rapid changes that characterize the knowledge-driven socioeconomic, political, and technological developments in the twenty-first century (Teichler, 2000). In the context of these changing conditions, higher education is faced with the task of reconceptualizing or reengineering its mission and roles. As declared by the participants of the Asia and Pacific Regional Conference on "National Strategies and Regional Co-operation for the 21st Century" (Regional Conference on Higher Education, Tokyo, Japan, 1997), and echoed by the World Conference on Higher Education (WCHE, 1998), the relevance of higher education institutions is shown by how their policies and practices coincide with the expectations of society.

Many studies on education and educational reform have been conducted to determine the responsiveness of the education system to the needs of Philippine society. Among these, the Congressional Committee on Education (EDCOM) Report that led to the promulgation of the Higher Education Act of 1994 has had a great impact on higher education. Under this Act, the Commission on Higher Education (henceforth referred to as CHED or the Commission) was created, thus limiting the focus of the Department of Education only to basic education.

The purpose of this chapter is to review the policies—*what ought to be*—and examine the practices and analyze problems—*what is*—relating to research in higher education in the Philippine context. To achieve this aim this chapter

- provides a background on higher education in the Philippines;
- presents the National Higher Education Research Agenda (NHERA, 1998–2007) developed by CHED and summarizes the Research Agenda

enumerated in the CHED Medium-Term Higher Education Development and Investment Plan 2001–2004;
- discusses the results of a survey on the status of research in Higher Education Institutions (HEIs) in the Philippines (1996–2001) and in the National Capital Region (NCR, 1996–2001), both conducted by the CHED Zonal Research Centre NCR II (Vicencio, Arciga et al., 2002b; Vicencio, Bualat et al., 2002);
- points out salient findings of surveys on the research capabilities of HEIs in the NCR (DLSU-Manila CHED Zonal Research Centre, 2002; Vicencio, Arciga et al., 2002a.)

The surveys of the NCR Zonal Research Centres were singled out in light of the fact that the greater number of research universities in the country are found in this region. The last section of the chapter focuses on the problems revealed by these survey results and offers recommendations—*what could be*—for addressing these issues.

Background on Higher Education in the Philippines

Distinctive Features, Facts, and Figures

Higher education in the Philippines is a dominant sector, with 2.5 million students enrolled in 1,605 tertiary institutions (as of January 2005). These institutions are distributed throughout the country, but the largest number (17.5 percent) is found in the National Capital Region. About 89 percent of these institutions are privately owned, relying mainly on tuition fees as their source of income. The remaining 11 percent are state/local universities and CHED-supervised colleges or institutions, which depend largely on government subsidy. In terms of enrollment, 70 percent of higher education students are registered in private institutions. This figure shows a substantial decrease from a share of 78 percent 10 years ago (academic year 1994–1995). In fact, there has been a continuing downward trend in the share of private institutions in total enrollments, from a high of 96 percent in 1955.

The Asian Development Bank (ADB) study on Higher Education in the Philippines (1999) notes that the Philippines is perhaps second only to the United States of America in terms of the number of higher education institutions. The population has a high degree of access to these institutions, with a ratio of 1 institution for every 66,000 people (compared to 1:500,000 in Australia and 1:166,000 in Indonesia). Moreover, there is a high "transition rate" between secondary and tertiary education, with about 90 percent of high school graduates moving on to postsecondary education. The survival rate (percent of cohort reaching fourth-year level), however, was only about 68 percent in the academic year 2001–2002, 3 percent lower than the rate 5 years earlier. Furthermore, the average number of graduates is 46,000 per year (De la Rosa, 2005).

CHED data indicate that business administration and related fields has consistently attracted the most number of students. In the academic year 2001–2002, this discipline group accounted for 26 percent of the students. The other popular programs are education and teacher training (18 percent), engineering and technology (15 percent), mathematics and computer science (11 percent), and medical and allied fields (7 percent). These data show the students' (and the institutions') preference for less expensive degree programs that are of low priority rather than the more expensive but more important ones for national development. The concentration of enrollment is in fields that are perceived to yield better job prospects for the graduates. In this regard, the biggest change was seen, on the one hand, in medical and allied fields, wherein the proportion of students dropped from 15 percent in 1994. On the other hand changes in the opposite direction were evident in the discipline groups of education and teacher training and mathematics and computer science (from 13 percent and 5 percent of the total enrollment, respectively, in 1994). These changes are attributable to corresponding adjustments in the labor market both locally and abroad.

In so far as the faculty is concerned CHED data for the academic year 2000–2001 show a predominance (59 percent) of bachelor's degree holders, with only 8 percent of the total 93,884 having doctorate degrees and 26 percent with master's level training.

Quality Assurance System

The primary reason for the creation of CHED in 1994 was to raise the quality of higher education through policy directions and a system of grants and incentives that put a premium on quality. The Commission therefore set forth the essentials of a Quality Assurance System (Internal and External), using *"peer evaluation systems"* (Padua, 1999). For Internal Quality Assurance, institutions planning to open a new program are required to accomplish a Self-Study Report (SSR), including a description of their Internal Control System. These SSRs are then audited or verified by an external Quality Assessment Team. Consequently, institutions that were granted a permit to operate a new program are subjected to periodic Performance Audits.

Other major reforms undertaken by CHED for quality improvement include operationalization of the technical panels for each major discipline, standards formulation, identification of "Centers of Excellence" and "Centers of Development," monitoring and evaluation, and support for higher education research (Valisno, 2000).

The accomplishments of CHED notwithstanding, it is perceived to be acting more as a regulatory agency rather than the development agency it was envisioned to be. Because of its "confused governance structure," it has been unable to provide strategic directions for itself and the system of higher education (ADB, 1999).

Accreditation System

One of the distinctive features of the Philippine education system is its functioning accreditation system. Accreditation is voluntary and is done for programs rather than institutions. An umbrella organization, the Federation of Accrediting Associations of the Philippines (FAAP), and its member agencies are recognized by the government through CHED. To date, 832 programs from 221 higher education institutions are in different stages of the accreditation process (Pijano, 2003). Although there is an increase of 11 percent over the total of 743 programs in the academic year 2000–2001, the number of accredited institutions represents only 14 percent of the higher education institutions (HEIs) nationwide.

The low percentage of institutions with accredited programs may be attributed to the difficult and demanding process of accreditation, in terms of the human and financial resources required. A growing concern about the varying standards of accrediting associations has also been noted, with the system for public institutions being perceived as applying rather low accreditation standards (Bernardo, 2003).

The relationship between the accrediting associations and the government is governed by the CHED memorandum order entitled "Policies of Voluntary Accreditation in Aid of Quality and Excellence," which specifies four levels of accreditation, and describes the criteria and benefits for each level. Institutions with Level 1 Programs are granted partial administrative deregulation. Those with Level 2 Programs are provided full administrative deregulation and partial curricular autonomy, as well as priority in funding assistance and subsidies for faculty development. Programs accredited at Level III receive full curricular deregulation and the privilege to offer distance education programs. Level IV institutions are given full autonomy from government supervision and control and eligibility for grants and subsidies from the Higher Education Development Fund (HEDF). Only one institution, De La Salle University, had reached Level IV status as of 2002. Since then, one more institution has been granted similar status.

Causes of Problems

De la Rosa (2005), until recently the chair of the Commission, identified four possible reasons to explain the rather negative picture of the state of higher education in the Philippines today: (i) lack of broad political and legislative support for real reform; (ii) unrestrained proliferation of state colleges and universities, local colleges and universities, and educational franchises; (iii) scarce budgetary allocation; and (iv) imbalance in student distribution.

The Role of Research in Higher Education in the Philippines

Higher education institutions (HEIs) have traditionally been concerned with the basic thrusts of teaching, research, and community service. These thrusts follow certain principles, spelled out by Mr. Federico Mayor, former UNESCO Director General, as follows: "to turn out responsible and committed citizens; provide the professional people needed by society; develop scientific and technical research; conserve and disseminate culture, while enriching it with contributions from each generation; act as the memory of the past and be alert to the future; and be a critical and neutral example which will be the vanguard of 'intellectual and moral solidarity' " (as cited in Ordoñez, 1997: 89).

Springing from these aims and principles are the tasks of producing skilled human resources for the marketplace, leaders in the political, social, and industrial/commercial fields, and research and creative works necessary for a nation's socioeconomic and political development (Tan, 1992).

Research, then, is an essential function of higher education institutions. Faculty members need a vast amount of knowledge to teach effectively. They also need to contribute to the production of knowledge for their disciplines to grow and progress.

Moreover, research should provide information that can lead to meaningful improvements in practice, policy, and reform in tertiary education. In the Philippines, in particular, research must contribute knowledge that can be helpful to national programs on sustainable development (Alcala, 1997).

Although acknowledged as a basic function, however, research is the most neglected task of HEIs in the Philippines. As observed by CHED (1997), the culture and environment for research leaves much to be desired. For this reason, the Commission drew up a National Higher Education Research Agenda (NHERA).

The National Higher Education Research Agenda (NIERA) (1998–2007)

The 10-year NHERA produced by CHED delineates the policies, priorities, strategies, procedures, and guidelines for the promotion, encouragement, and support of research in the public and private colleges and universities in the Philippines. This was done in view of CHED's mandate (The Higher Education Act of 1994, R.A. 7722) to

- formulate and recommend development plans, policies, priorities, programs, and research on/in higher education;
- recommend to the executive and legislative branches priorities and grants on higher education and research;
- develop criteria for allocating additional resources for research; and
- direct or redirect purposive research by institutions of higher learning to meet the needs of agro-industrialization and development.

Higher Education Research Goals

Recognizing the role that research plays in supporting instruction and extension activities in higher education institutions, the NHERA articulates the goals of higher education research as follows:

1. To push the frontiers of knowledge across all the identified higher education disciplines in the country.
2. To enhance instruction through original contributions in specialized disciplines, thereby encouraging students to become creative, innovative, and productive individuals.
3. To develop unifying theories or models that can be translated into mature technologies meant to improve the quality of life of the Filipinos within the sphere of influence of the academic institutions in the country.

Elements of the Higher Education Research Framework

The NHERA stipulates the mechanics and concrete steps to be taken for the achievement of the goals stated above.

General Policy Statements

The following general policies emphasize the development of a culture and environment for research in higher education institutions:

1. Research is the ultimate expression of an individual's innovative and creative powers. Higher education institutions shall ensure that the academic environment nurtures and supports Filipino research talents.
2. Research thrives in an environment characterized by free flow of information, honest and analytical exchange of ideas, and supportive administrative structures. Higher education policies shall enhance the individual's capacity to conduct independent and productive research.
3. Research is one of the main functions of higher education institutions. Universities, in particular, are expected to lead in the conduct of technology-directed and innovative/creative researches that are locally responsive and globally competitive.

These policy statements recognize that it is through individuals that tertiary institutions can carry out their research function. Research activities of college and university professors have traditionally been major sources of knowledge and innovation. Although students too, under the guidance of their professors, have the potential to contribute to the knowledge base, it has been observed that many of them do research merely to comply with school requirements. Many professors themselves do not conduct research beyond their master's thesis or doctoral dissertation. It is therefore important that the CHED stresses the need for higher education institutions to provide the environment that would encourage research efforts toward the improvement of the Filipinos' quality of life.

Support for Research

Policy directions of CHED encompass research management and administration, technical assistance programs for research, and funding for higher education research.

HEIs are expected to provide administrative support for research in the form of specific policies on deloading for research activities, the use of institutional facilities for research, and other such incentives. Both the CHED and the individual Higher Education Institutions (HEIs), on the other hand, are expected to offer technical and logistical support for research.

The Commission is committed to give technical help by making expertise (local and international) available to individual researchers in the various higher education disciplines. It also provides logistical and financial support by means of research block grants, research grants-in-aid, and grants for commissioned research.

Research block grants and grants-in-aid are offered to HEIs on a competitive basis. The awarding of block grants is based on the proposal's merit and impact on higher education. Only institutions that meet requirements concerning their faculty workforce, research structure, research facilities, and publication assurance may submit research proposals for these grants. Grants-in-aid, on the other hand, are given to developing HEIs that show potentials for improving their research capacities, are strategically located vis-à-vis government socioeconomic programs, and demonstrate willingness to provide counterpart funding for their research activities. Commissioned researches are awarded to academic institutions or individual recipients who have the suitable credentials for undertaking research projects deemed important for policy and decision making.

The Commission has allocated a total of PHP235.7 million (approximately US$4.3 million) from the Higher Education Development Fund (HEDF) for these research grants and aids (CHED, 2000a.)

Intervention Strategies
The CHED has identified several intervention measures to increase the quality and quantity of research outputs of HEIs. These include (i) the provision of technical and financial aid to HEIs that have research capability in critical disciplines; (ii) strengthening linkages with research institutions in Australia, the United States of America, and in European and Asian countries; (iii) financial support for externally refereed national research journals in various disciplines; (iv) the recognition of outstanding researchers via professorial chair awards and other such incentives; and (v) the development of high-level research-oriented human resources in critical disciplines through the use of the Commission's HEDF for the research training of promising junior faculty and graduate students.

In line with these strategies, the Commission has issued memoranda such as those providing guidelines for CHED dissertation grants (CHED, 2003b, Memorandum Order No. 04, Series of 2003), visiting research fellowships (CHED, 2003c, Memorandum Order No. 13, Series of 2003), and the "Republica Awards" for outstanding research and publications (CHED, 2003a, Memorandum from the CHED Executive Director, September 16, 2003).

Priority Research Areas

General Principles
The choice of priority research areas for the National Higher Education Research Agenda was guided by the principles of multidisciplinarity, policy orientation, operationalization, and participation/broad impact. Preference is shown for studies that involve research experts from several disciplines, are policy-oriented, investigate and explain the relationship of different phenomena or factors, involve a good number of stakeholders, and have impact on a wide range of individuals or groups. "Breakthrough" or pioneering researches, which need not be multidisciplinary in nature, are also favored.

Priority Thrusts
Following the general guidelines, broad priority research areas have been identified in the following disciplines: science and mathematics; engineering, maritime studies, and architecture; humanities, social science and communication; agriculture and fisheries; business and industry; health and health-related disciplines; information technology; teacher education; and industrial technology.

The multidisciplinary researches suggested in these disciplines are aimed at advancing the frontiers of knowledge in these areas or improving the concepts, products, or services related to these areas.

Other emphases in higher education research are specified:

1. Program/curricular assessment studies on higher education clusters of disciplines.
2. Research on integrative theories, models, or philosophy; policy-oriented studies on financing of higher education, economics of higher education, governance and management of higher education, accreditation of HEIs, rationalization of higher education.
3. Model-building and institution-building studies.
4. Labor market supply and demand studies.
5. Integrative studies in linguistics, sociology, anthropology, and other social sciences.
6. Other topics responsive to emerging needs of the country (i.e., socially oriented and community based studies).

Zonal Research Centres (ZRCs)

Twelve Zonal Research Centres (ZRCs) were established in 2000 to help the Commission in efficiently and effectively managing research activities of higher education institutions. The duties and responsibilities of these ZRCs within their respective region(s) of jurisdiction include

- serving as a clearinghouse of research proposals submitted by Higher Education Institutions (HEIs) for CHED-HEDF funding;
- assisting the Commission in the monitoring of CHED-funded researches, conducting training programs to build the research capabilities of the HEIs;
- undertaking at least one commissioned research as identified by CHED each year; gathering, soliciting, and screening research outputs of HEIs for possible publication in *CHED's Philippine Journal of Higher Education*;
- serving as a guiding force between and among higher education institutions (HEIs) relative to their research networking activities; and
- administering the funds provided (PHP1million [about US$18,000] each, per CHED Resolution No. R-57–2000) for initial operating costs (as stated in CHED, 2000c, M.O. No. 08, Series of 2000).

Medium-Term Higher Education Development and Investment Plan 2001–2004: Research Agenda

CHED formulated a Medium-Term Higher Education Development and Investment Plan (2001–2004) 3 years after the establishment of the National Higher Education Research Agenda. The Plan "provides the policy framework and defines the programmes that will enable the higher education sub-sector to fulfil its role in the development of the country's human resource in the context of globalization and the emerging knowledge-based economy."

As part of its strategies for promoting relevance and responsiveness, the Commission stated the goals of "ensuring labour market responsiveness of higher education and strengthening the research and extension functions of HEIs" (CHED, 2000b, Medium-Term Development Plan). In this light, a research agenda was developed, which includes

- rationalization studies;
- benchmarking and comparative study of policies, standards, and guidelines in various disciplines in Asia, Europe, and the United States of America;
- establishment of quality indicators;
- impact study on liberalizing entry of foreign universities/colleges via satellite;
- evaluation of graduate programs in teacher education and business education;
- graduate tracer studies; and
- impact study of ICT-driven curricula on student learning and academic performance. These topics are meant to operationalize the priority thrusts identified in the National Higher Education Research Agenda (NHERA).

The Status of Research in Philippine Higher Education Institutions (HEIs)

Three years after the National Higher Education Research Agenda (NHERA) was developed, and 1 year after the establishment of the Zonal Research Centres (ZRCs), a study was conducted to determine the status of research in HEIs in the Philippines (Vicencio, et al., December 2002).

Quantity Generated

The documentary survey covered 13,859 research reports submitted by 259 respondent tertiary institutions to the 12 ZRCs all over the country. These were studies conducted within the period 1996–2001. The low turnout (a mean of 10.71 studies/HEI/year) and inadequacy of data processed were attributed to the HEIs' lack of an efficient system of monitoring research outputs (Vicencio, Arciga, et al., December 2002).

The survey revealed that 60.33 percent of these studies came from institutions in the National Capital Region (NCR). Region 7 accounted for the next highest number of researches (11.58 percent), and the rest of the regions contributed anything from 0.89 per cent to 6.39 percent.

The large number of studies produced in the NCR is consonant with related findings of the two Zonal Research Centres (ZRCs) in NCR based on the research capabilities of Higher Education Institutes (HEIs) under their jurisdiction. All the 23 respondent HEIs in NCR Group 2 and 77 percent of the 61 respondent HEIs in NCR Group 1 affirmed that they undertake research and development activities

(Vicencio, Bualat, et al., 2002; DLSU-Manila CHED Zonal Research Centre, 2002). These activities include personal grants and institutional awards. It was noted, however, that only 68 percent of the faculty in the HEIs under the jurisdiction of ZRC 2 were involved in research.

A review of the completion dates of these studies, 1 year after the introduction of the NHERA, shows that 1999 was the most productive year, with 24.29 percent reported for this period (compared to the average of 16.46 percent from the previous 3 years). The productivity was not sustained thereafter (average of 13.16 percent in the next 2 years), however, raising doubts about the possible attribution of the surge in researches in 1999 to the NHERA. Reasons offered for the downtrend after 1999 were the discontinuation of the grants-in-aid for research in 2000, the delayed documentation of projects conducted from 2000 to 2001, and difficulties in record keeping and development of databases (Vicencio, Arciga, et al., 2002). This national profile was reinforced in the study limited to institutions in the NCR Group 2 (Vicencio, Bualat, et al., 2002).

Types of Research/Researchers

Among the 13,859 studies reported for the 6-year period, researches conducted by individuals (72 percent) far outnumbered collaborative or institutional researches. This result is consistent with findings of earlier reviews of higher education research in the Philippines (Bernardo and Sarmiento, 1997) and of studies done in NCR Group 2 alone (Vicencio, Bualat, et al., 2002). However, in contrast to the national picture showing master's theses and doctoral dissertations constituting about 69 percent and individual faculty and staff researches constituting 31 percent of these individual studies, the NCR Group 2 review showed slightly more individual faculty/staff researches (52.90 percent) than graduate students' researches (47.10 percent). The 2002 national picture more closely resembles previous survey results (Bernardo and Sarmiento, 1997) that showed close to 60 percent of researches having been done by graduate students (master's and doctoral) as part of their degree requirements. This implies that the studies were one-shot short-term projects that neither built on earlier findings nor led to further investigations. Needless to say the utilization of knowledge derived from these researches is therefore rather limited.

Majority of the studies (56.81 percent) employed the descriptive research design, with survey questionnaires as the most popular data collection tool (used by 30.93 percent). A similar trend was seen in the NCR 2 data (Vicencio, Bualat, et al., 2002) and in Bernardo and Sarmiento's (1997) review. These and the earlier observations about the authorship of the studies surveyed seem to indicate that the NHERA priorities of "leading-edge" scientific or technological research or of multidisciplinary research have not been followed.

Research Thrusts

Studies on the research capability of HEIs in the NCR Groups 1 and 2 (DLSU-Manila CHED Zonal Research Centre, 2002; Vicencio, Bualat, *et al.*, 2002) reveal that the greatest number of researches done was in the field of education and teacher training. This is not surprising, considering that most graduate programs are in this area. Humanities, social/behavioral sciences, mathematics, and computer science were the next most productive disciplines. In addition to these, business administration, medical and allied sciences, and natural sciences attracted researchers in at least 50 percent of the institutions in NCR Group 2. The least amount of researches generated was in the fields of law and industrial technology for the 61 HEIs in NCR Group 1, and in aeronautics/aviation, maritime education and training, music and dance, economics, and 5 other fields for the 22 institutions in NCR Group 2. The NCR Group 2 study further indicates emphases on program/curricular assessment, institution-building studies, integrative studies in linguistics, sociology, anthropology, the social sciences and humanities, as well as policy-oriented studies.

A review of the titles of 42 research proposals approved for funding by CHED ZRC NCR Group 2 (2002) shows that 26 percent are in the natural sciences and 19 percent in the social and behavioral sciences. Other areas covered (with 10–12 percent each) are educational and teacher training, architecture, trade craft and industry, and mathematics and computer science. In an earlier report of CHED (2000a), the largest number of ongoing, completed, or approved-for-funding projects was also in the natural sciences (33 percent of 15 applications). Trade craft and industry as well as educational and teacher training had the next highest frequency of projects (27 percent and 20 percent, respectively).

These findings suggest that the researches conducted in the NCR HEIs follow, for the most part, the thrusts identified in the NHERA. The following disciplines do not get enough research attention: engineering, maritime studies and architecture, agriculture and fisheries, and industrial technology. Moreover, there is a need to put more emphases on other priority areas in higher education listed in the NHERA, such as integrative theories, models, or philosophy; financing and economics of higher education; governance and management of higher education. Furthermore, the choice of research topics ought to reflect more clearly the principles of multidisciplinarity and broad impact.

Practices

Incentives
Most of the institutions in the NCR surveys (DLSU-Manila CHED Zonal Research Centre, 2002; Vicencio, Arciga, et al., 2002a) indicated that they provide incentive programs for their researchers. These incentives may take the form of attendance in local/international conferences, honoraria, awards, promotion schemes, publication of research outputs, research load credit, merit points, and granting of sabbatical leaves. A good number of HEIs in NCR 2 (60.85 percent) also assist researchers by way of purchasing needed equipment for research. That only about 53 percent of the faculty members are involved in research despite these forms of administrative support probably indicates that such extrinsic rewards may not be sufficient to motivate faculty to do research. The need for the concomitant technical support is evident from the finding that few institutions (13.04 percent or less) offer technical assistance or conduct in-house training and capability-building seminars. Moreover, logistical and financial support are very much needed, considering that only 59 percent of the NCR Group 1 institutions have budget allotments for research and that only 1–10 percent of the NCR Group 2 institutions' budget are allocated for research. Although

many HEIs have research offices/units and research policies, many more lack research facilities, particularly research library resources.

Funding

In view of the meager budget allocation for research, HEIs have to solicit research funds from other sources. The NCR surveys show that those who do seek such assistance most of the time approach government sources, nongovernmental organizations, and private funding agencies. For its part, CHED (2000) reported having funded 16 research projects with a total approved budget of about PHP9 million.

Research Dissemination

For the HEIs in the NCR, local (i.e., institution-based) publications and forums are the most popular (cited by 38–46 percent of the HEIs) modes of disseminating research results. International and national forums as well as national and international publications are the next most frequently cited, although to a much lesser degree (16–30 percent). These findings imply that the information generated by higher education research is largely confined to local institutions and would therefore be unable to contribute much to building a coherent body of knowledge or create an impact on the wider community. The publication of research reports in externally refereed journals should therefore be encouraged.

Problems Identified and Recommendations for Addressing These Issues

The current status of higher education research in the Philippines, as gleaned from the surveys discussed above, does not show much improvement from the state portrayed by the Congressional Committee on Education (EDCOM) in 1993. At that time, EDCOM reported that the research produced by HEIs were "repetitive and stereotyped" and leaning heavily on the field of education and allied fields, with the sciences given low priority. EDCOM further described the *quality of research outputs as below world standards* and noted the lack of studies dealing with the development of unifying theories and models or new programs and strategies.

Consultative meetings conducted by CHED (1997) identified the following reasons for the poor research performance of HEIs: "inadequate public education, information and campaign on research results; low rate of public investments in research and development; inadequate allocation of funds; weak coordination among higher education institutions; inadequate or lack of research facilities and library resources and other logistics to support research; and the conduct of research by students merely to comply with school requirements."

It is evident that the problems of research in higher education in the country revolve mainly around the research capability of the institutions. This refers to the university infrastructure that supports the students, faculty, research programs and centers, including the provision of facilities, technical and logistical support, research training, and research management policies. Underlying all of these is the need for financial resources.

As stated earlier, about 89 percent of 1,605 higher education institutions in the Philippines rely mainly on tuition fees as their source of income, and the remaining

11 percent depend largely on government subsidy. As more and more public HEIs are established, however, their share of the public funds becomes less and less. Under these conditions for both the private and the public institutions, economic constraints and changing priorities usually result in research receiving little support. External funding is therefore needed to promote and sustain the institutional capacity of these higher education institutions (HEIs).

The fact that CHED has allotted PHP235.7 million for research grants and aids is certainly a positive development. But this translates to meager amounts allocated for approved projects, necessitating additional support from the institutions themselves, which are treated uniformly by CHED. As observed by Bernardo (2003), there is a need for the Commission to allocate more strategically the limited research development funds.

The Commission can further assist HEIs by way of advocating research support from private funding agencies. In this regard, researchers would have better chances of obtaining external funding if they were able to show clearer connections between their proposed research and the expected value of research results in the solution to specific social problems. This implies that researchers would have to go beyond the usually narrow, limited focus of their one-shot research studies, begin to explore major concerns and issues, and seek innovations to address these.

This point leads to another central problem of higher education research in the Philippines, namely, its *human capital*. The minimal involvement of faculty in research activities may be better understood in view of the fact that many of the HEIs are purely teaching institutions, a number of which are secondary schools that have been upgraded to the tertiary level. Furthermore, the CHED data for the academic year 2000–2001, which show that a bachelor's degree is the highest educational attainment for about 59 percent of faculty members in Philippine HEIs, would indicate that many of the faculty lack the training and experience to do research. Moreover, among those who have obtained graduate degrees, few have done research beyond their master's theses or doctoral dissertations. If these data are juxtaposed with the recent findings of the CHED-commissioned Evaluation of Graduate Education Programmes (CHED-FAPE, 2005), which rated as *Poor* 13 percent of the graduate programs in teacher education, 7 percent in business education, and 22 percent in public administration, even the quality of faculty who may have completed post- baccalaureate degrees from these institutions becomes suspect. Human resource development is therefore a major area for improvement in higher education institutions for the advancement of research.

The nexus between teaching and research is clearly missing in tertiary institutions that have not invested in research. This suggests a need to clarify the goals and purposes of higher education. HEIs that choose to remain merely teaching institutions need to realize that good teaching can only be maintained if teachers are active researchers. As Boyer (1990) indicates, the teaching and research functions involve an interlocking of four activities of scholarship: the scholarship of discovery, of application, of integration, and of teaching and learning. Each HEI must, therefore, try to identify its own mission within that spectrum. In the end, some may be expected to engage in research to a greater extent than others; but none can totally avoid such involvement.

Considering that graduate schools are the major source of researches in the Philippines, there is a need to look more closely at their research outputs. The proliferation of graduate programs in teacher education explains the huge number of studies in this area, to the neglect of others. The descriptive nature of most of these studies, many of them conducted in the researchers' own institution out of convenience, puts into question their technical adequacy and social/professional relevance. The results of the EGEP project could therefore lead to crucial decisions on the continued existence of graduate institutions offering substandard quality of teacher education, business education, and public administration programs, as well as to reforms in all aspects of graduate education, especially research.

The difficulty in collecting research information from the higher education institutions brings out another problem—the lack of an adequate database. Efforts on the part of these HEIs and of CHED to monitor research activities more closely and develop an electronic database can help researchers avoid reinventing the wheel and determine what frontiers have yet to be explored.

The state of higher education research in the Philippines, even after the development of the NHERA, reveals a great need for a stronger push from all sides. The stakeholders—higher education institutions, the CHED, other government agencies, private funding agencies, and industry—must join hands to achieve the goals of higher education research. This should be done within the framework of the expanded vision of the core mission of higher education, as proposed by the World Conference on Higher Education (WCHE, 1998), that focuses on "equity of access, increased participation of women, the advancement of knowledge through research and dissemination, and the need for increased emphasis on relevance, closer ties with the world of work and anticipation of societal needs" (UNESCO Asia and Pacific Regional Bureau for Education, 2002). In the context of technological advances in a globalized world, this implies adherence to Boyer's new paradigm of scholarship (Boyer, 1990; Glasick et al., 1997).

Acknowledgment

This chapter is based on papers presented by the author at the 1st Regional Research Seminar for Asia and the Pacific of the UNESCO Forum on Higher Education, Knowledge and Research, May 13–14, 2004 United Nations University, Tokyo, Japan; at the Asia and the Pacific Conference on Higher Education Research, August 18–20, 2004 Manila, Philippines; and at the UNESCO Regional Seminar on the Implication of WTO/GATS on Higher Education in Asia and the Pacific, April 27–29, 2005 Seoul, the Republic of Korea.

References and Works Consulted

ADB (Asian Development Bank) (1999) "Higher Education in the Philippines." Part 1, *The Philippine System of Higher Education*. Available from the World Wide Web: http://www.adb.org/ Documents/Books/Phil-Education/Higher Education in the Philippines.pdf (accessed July 31, 2005).

Alcala, A. (1997) "Foreword." In the Commission on Higher Education, ed., *The National Higher Education Research Agenda (1998–2007)*. Report. Pasig City: Commission on Higher Education.

Bernardo, A. B. I. (2003) "International Higher Education: Models, Conditions and Issues." In T. S. Tullao, Jr., ed., *Education & Globalization*. Philippines: Philippine APEC Study Centre Network (PASCN) and the Philippine Institute for Development Studies (PIDS).

Bernardo, A. B. I., and J. E. Sarmiento (1997) "Toward the Rationalization of Research on Higher Education: A Survey of Higher Education Research in the Philippines (1975–1996)." Edukasyon. UP-ERP Monograph Series 3, January–March 1997.

Boyer, E. (1990) *Scholarship Reconsidered: Priorities of the Professoriate*. Princeton, NJ: Carnegie Foundation for the Advancement of Teaching.

CHED (Commission on Higher Education) (1997) "The National Higher Education Research Agenda (1998–2007)." Report. Pasig City: Commission on Higher Education, 3.

—— (2000a) "CHED Accomplishment Report of FY 2000." http://www.ched.gov.ph/aboutus/acc_report_2000.html (accessed July 31, 2005).

—— (2000b) "Medium-Term Higher Education Development Plan (2001–2004)." http://www.ched.gov.ph/about us/medterm_plan.html (accessed July 31, 2005).

—— (2000c) Memorandum Order No. 08, Series of 2000.

—— (2003a) "Memorandum from the Executive Director" on Guidelines for CHED Republica Awards, September 16, 2003, Pasrg City.

—— (2003b) Memorandum Order No. 04, Series of 2003.

—— (2003c) Memorandum Order No. 13, Series of 2003.

CHED-FAPE (2005) "Report on the Evaluation of Graduate Education Programmes (EGEP)." Makati City, Philippines, Fund for Assistance to Private Education, August.

De la Rosa, R. V. (2005) "Towards a Deeper Understanding of the Commission on Higher Education and Our Role as Educators." Keynote address delivered at the CHED 2nd Regional Consultative Conference, Makati City, Philippines, April 12, 2005 (unpublished manuscript).

DLSU-Manila CHED Zonal Research Centre (2002) "Research Capabilities of HEIs Under NCR Group I." Unpublished report submitted to the Commission on Higher Education.

EDCOM (Congressional Committee on Education) (1993) *Tertiary Education*. Book 2, vol. 3. Quezon City: Congressional Oversight Committee on Education.

Glasick, C. E., M. T. Huber, and G. I. Maeroff (1997) *Scholarship Assessed: Evaluation of the Professoriate*. San Francisco, CA: Jossey-Bass.

Ordoñez, V. M. (1997) "UNESCO's Vision for Higher Education and Human Resource Development for the 21st Century." In *Proceedings of the World Congress on Higher Education* (June 23–25, 1997). Manila: CHED, 87–99.

Padua, R. N. (1999) "Quality Assurance in Philippine Higher Education." In "Quality Assurance for Higher Education." A report from the Regional Workshop on Quality Assurance for Higher Education. Bangkok, Thailand, November 24–26, 1998, Bangkok, Thailand, SEAMEO Regional Centre for Higher Education and Development.

Pijano, C. V. (2003) "Import and Export of Higher Education: How to Sustain Quality: Experience in the Philippines." Paper presented at the International Network for Quality Assurance Agencies in Higher Education (INQAAHE), Asia Pacific Quality Sub-Network Forum, Hong Kong, China, January 17–18, 2003.

Regional Conference on Higher Education (1997) "National Strategies and Regional Co-operation for the 21st Century." Declaration concerning higher education in Asia Pacific Region, Tokyo, Japan, July 8–10, 1997.

Tan, E. A. (1992) "State of Higher Education in the Philippines." *Ugnayan: Philippine Journal of Higher Education*, 1(1):6–28.

Teichler, U. (2000) "The Relationships between Higher Education Research and Higher Education Policy and Practice: The Researchers' Perspective." In U. Teichler and J. Sadlak, eds., *Higher Education Research: Its Relationship to Policy and Practice*. Kidlington, Oxford: International Association of Universities and Elsevier Science, 3–34.

UNESCO Asia and Pacific Regional Bureau for Education (2002) "Higher Education in Asia and the Pacific." Draft Regional Report on progress in implementing the recommendations of the 1998 World Conference on Higher Education (WCHE), Bangkok, Thailand, UNESCO, December 13, 2002.

Valisno, M. (2000) "Quality Assurance in Philippine Higher Education: Lessons Learned." In G. Harman, ed., *Quality Assurance in Higher Education. Proceedings of the International Conference on Quality Assurance in Higher Education: Standards, Mechanisms and Mutual Recognition*, Bangkok, Thailand, November 8–10. Bangkok, Thailand: Ministry of University Affairs of Thailand and UNESCO Principal Regional Office for Asia and the Pacific.

Vicencio, E. M., A. J. P. Arciga, A. S. Bualat, P. A. C. Famularcano, and A. P. De la Cruz (2002a) "Research Capability of Higher Education Institutions in the National Capital Region II." Unpublished report submitted to the Commission on Higher Education, October 2002.

—— (2002b) "Status of Research and Development in Higher Education Institutions in the National Capital Region II (1996–2001)." Unpublished report submitted to the Commission on Higher Education, December 2002.

Vicencio, E. M., A. S. Bualat, P. A. C. Famularcano, and A. P. De la Cruz (2002) "Status of Research and Development in Higher Education Institutions in the Philippines (1996–2001)." Unpublished report submitted to the Commission on Higher Education, December 2002.

WCHE (World Conference on Higher Education) (1998) "World Declaration on Higher Education for the 21st Century: Vision and Action." Paris: UNESCO, October 5–9 1998.

CHAPTER ELEVEN
HIGHER EDUCATION REFORM IN THAILAND

Charas Suwanwela

Introduction

The population of Thailand for the year 2001 was 62.6 million; following the baby boom of the mid-1970s the annual population growth rate totaled 3.2 percent. The adult literacy rate for the same year amounted to 86.8 percent—an increase of 84.4 percent from 1997. However, due to the successful "Family Planning Programme" the annual population growth rate dropped to 0.7 percent from the year 1997.

The population cohorts for those born during the height of population growth, which was about 1 million annually, have just passed the 18–24 age groups, with a gradual drop in the number of children attending primary education in the past decade. Fifty years ago compulsory education for all citizens was raised from 4 to 6 years, then 9 years, and since year 2000, 12 years of study; the result is that a substantial percentage of students, graduating from secondary schools, are seeking higher education. The majority of the workforce has shifted, to a great extent, from the agricultural to the industrial sector, creating great demand for vocational, professional, and continuing education. There is also a rapid increase in the percentage of senior citizens (Office of the National Education Commission, 2000).

Politically speaking, Thailand is now one of the most stable countries in the region. A coup d'état in 1932 resulted in a constitutional monarchy replacing the absolute monarchy, and Thailand became a democratic region. In the early years the elected parliamentary democratic system was periodically interrupted by "military dictatorship." Following the revolutionary overturn in 1992 of a military government the people have enjoyed a new constitution with a more democratic system which guarantees basic rights for all citizens, including the right to education.

Economically speaking, the country went through periods of difficulty, such as during the oil crisis in the early 1980s, followed by the East Asian economic collapse in 1996. However, in between there were periods of remarkable economic growth, especially during the early 1990s.

As from 1996, the economic and social crisis persisted for a number of years followed by gradual recovery. But unfortunately the higher education system could not cope with the changes; during the "economic booms," there were shortages of trained manpower, while during the crisis, unemployment and underemployment of university graduates prevailed. In the years of crisis, higher education programs also served to divert unemployed university graduates to further university education. The gross domestic product (GDP) per capita during the 1997 "soap-bubble economic expansion"—Thailand shifted to the managed float system—was US$3,035, but this dropped to US$1,831 in 2001.

Universities for "Knowledge Transfer"

Universities in Thailand have a reputation for "knowledge transfer"; their main function is teaching. The higher education system in Thailand came about following the creation of a Law School in the Ministry of Justice in 1887, soon followed by a medical school, the Royal Pages School, for training in government administration—and an engineering school. These schools were combined to form one university in 1917, which represented Thailand's first university. Higher education aimed at training prospective civil servants to serve the needs of modernized bureaucracy and infrastructures such as railroad, postal service, irrigation, and healthcare.

It was conceived worldwide that there was a large quantity of knowledge transfer in European countries and that higher education institutions would serve to channel it in order to benefit the teaching and production of professionals and the education system of Thailand as a whole. It is interesting to compare this with the development in Japan, which began more or less at the same time; however, Japan followed the German model where the university was for research: the "pursuance of truth and knowledge transfer."

Up until 1934 Chulalongkorn University remained the first and only university in the country, when at that period the Thammasart University was established. The Thammasart University was founded soon after the Democratic Revolution in 1932 with the aim of educating a greater number of people in the moral and political sciences.

The quantitative expansion, during the 1960s and 1970s, with the creation of more universities both in the capital city and in the provinces, as well as vocational, agricultural, and teacher-training colleges was compatible with the National Economic and Social Development Plan (NESBD). Institutions were established following the lines of government bureaucracy as the main objective of the universities was to "serve" the civil services.

There was severe economic depression during World War II resulting in the government issuing banknotes, the result of which was chronic inflation. Civil servants salaries, which included those for university teachers, were allowed to deteriorate. With a bigger workload from an increasing number of students—plus the time lost from seeking additional income outside the universities—the quality of teaching was compromised and research was at its minimum. Academic progress was slow and was put on the shoulders of the new up-and-coming generations of young academics who were sent to study in Western countries. Upon their return to Thailand, they were overwhelmed by the teaching loads although some found the time to produce knowledge through research but unfortunately without much support from the powers in office.

Quantitative Expansion of Universities

Political pressure, population growth, and socioeconomic development led to the quantitative expansion of higher education services. New public universities and colleges were established with inadequate investment. Following several student uprisings there was increased demand for higher education. In 1971 an Open University was founded, and in 1981 another one based upon distance education was created. Provision of higher education by private sector and foreigners has been allowed since 1965. At the very beginning colleges were established; the Private College Act of 1969 enabled the elevation of private schools and colleges to degree-granting level. By 1984, four degree-granting private universities were in operation. Since then there has been very rapid expansion. The rate of growth has accelerated over the past 3 decades, as evidenced by the increase in the number of institutions and the huge quantities of enrollments.

Student enrollment in the higher education system went up a mere 69,000 in 1970 to almost 800,000 students in 1984. There were, however, a disproportionate number of students enrolled in the social sciences, an area that lent itself more easily to rapid expansion than did the natural sciences. The higher education institutions limited their functions to teaching. As a result of the response to demand for enrollment expansion followed in unemployment and underemployment of university graduates. The situation deteriorated again in the early 1980s due to the nation's economic difficulties that stemmed partly from the oil crisis.

However, during the second half of the 1980s the situation reversed itself. The nation's economic growth soared remarkably due to an inflow of foreign investments and industrialization; manpower shortages of scientists, business administrators, and in particular engineers were keenly pronounced.

Response to the need to produce more graduates in specific fields was constrained by the loss of faculty members to private industries and businesses. The problem of overproduction of school teachers was further aggravated by the shrinkage in the number of primary school students due to the successful family planning efforts. As a result, many teacher-training colleges in the provinces became community colleges and universities. The number of higher educational institutions was 43 (26 private) in 1990, 66 (42 private) in 1998, and 78 (54 private) in 2002. The increase in demand stemmed both from the larger number of those finishing secondary schools and a larger proportion seeking higher education. For instance, in 1988, 244,034 students graduated from secondary schools and 144,931 or 59 percent entered into higher education; while in 1998, 480,609 finished secondary schools and 364,871 or 76 percent entered colleges and universities (Office of the National Education Commission, 2000).

The expansion responded to youth's demand, which did not correspond to the world of work, resulting in unemployed graduates and at the same time shortages in certain disciplines. There has been an imbalance between science and technology on the one hand, and social sciences and humanities on the other, the latter of which are subjects that can be offered with less investment cost. The overall ratio was 22:78 in 1998. For public universities not including open universities the ratio was 56:44 indicating the attempt by the public sector to address this issue; the open universities had a ratio of 8:92 while it was 19:81 for private institutions. With many new universities in the provinces, the situation is likely to worsen.

Equity in higher education has always been a major concern in Thailand. In spite of the large expansion of the higher education system access to higher education based on merit remains unequal. The difference in quality of secondary schools in the country is big. Gaps exist between better public and private schools in big cities and those in rural areas. The distribution of higher education institutions is also a problem; political and other influences have dictated the location of new universities in the past. National entrance examination and matching systems that have been in place for 4 decades have prevented the use of influences and corruption in the process, but it has created fierce competition and the diversion of students' interest from regular schooling to tutorial in preparation for the examination. The quota system for students in the region where the universities are located has helped to provide better chances for students in the provinces. The admission system is currently under review.

In 1995, the government established a loan scheme for needy students for families with an annual income below THB300,000. The limit was later lowered to THB150,000. Between 1996 and 2001 over 200,000 students benefited from more than THB12 billion due to this scheme.

Relatively low investment in higher education that has prevailed for several decades has been one of the root causes of problems, especially with regard to quality. With rapid massification the situation could be expected to get worse (Suwanwela, 2002b).

In 2002, there were over 1.7 million students enrolled in 126 higher education institutions. About 1 million students were in 24 public universities, 86 percent sitting for a bachelor degree. Among them, 660,000 were enrolled at open universities. There were about 220,000 students in 56 private universities and colleges with 95 percent at the bachelor level. The male/female ratio for all students was 41:59; girls were more successful at the entrance examination. The percentage of students in social sciences and humanities was 77 percent overall, but the figure was 94 percent for open universities. While 96 percent of students in regular universities completed their study and graduated, only 26 percent of those enrolled at open universities got their degree. A 2001 survey of university graduates revealed that 35 percent were still unemployed 1 year after graduation.

Research in Thai Universities

As early as 1929, Prince Mahidol reported to the government that one function of the university was missing and that was the *"inquiry and pursuance of knowledge"* and consequently the Rockefeller Foundation (RF) gave assistance to upgrade the medical school. Professors from Western countries were recruited to teach in basic and clinical sciences, while Thai students were sent to study abroad. Upon their return, they became responsible for teaching, but some were competent enough to carry out small research projects.

In 1959, the First National Economic Development Plan (2001) was established, and the need for research in the country was recognized. The National Research Council (NRC) was founded with ten committee sectors to draw up plans and initiate financial support for research. Through the council, research grants were given to university teachers and government departments. Government ministries including the Ministry of University Affairs (MUA) could, however, seek their own budgets for research. The council only benefited from a very small fraction of the funds available. This resulted in an imbalance of the distribution of the research budget; with a large portion going to agriculture. Gradually the research efforts shifted away from the national development directions. There was also a fragmentation of research efforts in various departments, institutes, and centers of many ministries; but an attempt to encourage the NRC to screen all research projects to be supported by the government was ineffective.

In early 1970 voices were heard from academic meetings and faculty senates regarding the need to accelerate research activities at Thai universities. Research institutes were established inside both universities and government departments. The criterion for academic promotion, which was overseen by the MUA, was altered to include research publications. The goals of universities were aimed at research for the "production of knowledge" as an educational tool especially for expanding graduate education. Some universities set the research university as their main goal. Offices of research affairs were set up in the universities and some had a vice-president for research affairs. Research planning, support, and coordination led to a number of large projects and programs. Local and indigenous problems such as tropical diseases, Thai and Southeast Asian languages and area studies, Thai natural resources, Thai architecture, as well as unique social problems were addressed. Unfortunately the basic sciences in general lagged behind (Suwanwela, 2002c).

There was also a corresponding expansion of graduate education toward master and doctoral degrees. It covered a sporadic 40 years but expansion was extremely slow at the beginning. There were only 1,434 students at the graduate levels in 1994; while in 2001 there were 43,238 master degree students and 2,441 doctoral degree candidates (Ministry of University Affairs, 1991).

In 1993, laws were passed to established three independent agencies for the promotion of research: (i) Thailand Research Fund (TRF); (ii) National Science and Technology Development Agency (NSTDA); and (iii) Institute of Health Services Research (IHSR). The TRF was given a budget as an endowment to promote "research through management" and "research grants." In its 10 years of existence, it has developed a management system for program planning, research support, and operation oversight. Senior researchers have been recognized and supported to form research teams; young researchers are supported through postdoctoral fellowships and research grants. A certain number were recruited to function as research managers. A special program for support of research in conjunction with doctoral training has resulted in attracting more than 1,000 students to doctoral education, the largest number ever heard of in Thailand (Thailand Research Fund, 2000).

The NSTDA offers both intramural and extramural programs; the former are in the three centers under the jurisdiction of the agency, the National Electronic and Communication Technology Centre, the National Biotechnology Centre, and the National Metallurgy and Material Centre. Besides the setting up of national plans, the agency also constructs and operates the first Thailand Science Park. Research grants have been given to support scientific research and innovation both to intramural centers and to universities (National Science and Technology Development Agency, 1998).

The Institute of Health Systems Research has recruited researchers from health and social sciences to conduct research in areas of national concern. It has been instrumental in the launching of a national policy on universal coverage for health care, and a number of other initiatives (Institute of Health Systems Research Annual Reports 1994, 2000, 2003).

An analysis of the research situation during the 7th National Social and Economic Development Plan, 1992–1996 (Areegul, 2000) revealed that the government budget for research was THB23 billion for the 5 years (1992–1996); 0.72 percent of the total budget (see table 11.1). It was 0.12 percent of GDP and low compared to the neighboring countries as follows: 0.34 for Malaysia, 1.1 for Singapore, 2.8 for Korea, and 2.45 for Japan. Even though Thailand had a national plan to enhance industrial development during these 5 years, 56 percent research investment by the government, amounting to THB3 billion, went to agriculture and agricultural industries and covered mainly 4 departments in the Ministry of Agriculture. Universities received only 6.5 percent of the 5 years' spending allocation; of an amount totaling some THB1,600 million, social sciences and humanities were allocated 53 percent, while science and technology received the remaining 47 percent.

During the first 4 years following its establishment in 1993, the National Science and Technology Development Agency (NSTDA) was allocated a budget of THB2,685 million. NSTDA was instrumental in bringing to the fore research and development on the following three selected fronts: (i) electronic and communication technology; (ii) material technology; and (iii) biotechnology.

The Thailand Research Fund (TRF) received an initial endowment of THB1,200 million, but for the following years the annual figure was reduced to THB100–300 million rather than the projected THB1,000 million a year over 10 years. However, the NSTDA had been given completed autonomy to manage the fund and "stimulated" research activities and graduate education in universities in all disciplinary and interdisciplinary areas. Initial support to basic research was 44 percent of the total expenditure with the social sciences being significantly represented.

Table 11.1 The budget for research outlined in the 7th National Development Plan

Research Field	Budget in billion THB.	Percentages (%)
Sciences/Technology/Industry	3,432	14.56
Agriculture that includes agricultural industry	13,454	57.06
Health	811	3.44
Social/Culture	1,362	5.78
Others	4,517	19.16
Total	23,576	100.00

Source: National Science and Technology Development Agency (NSTDA), 1998.

Major Reform Since 1997

The promulgation of the 1997 New Constitution in Thailand, and the passing of a New Education Act two years later have initiated major reform. This New Constitution of 1997, states clearly the "right to education for all citizens" as well as the "duty of the state to provide free 12-year compulsory education." It is interesting to note that it also is necessary for candidates who wish to become "elected members of Parliament" and "cabinet ministers" to have, at least, a bachelor's degree. The National Education Act B.E. 2542, outlined some major structural changes as follows: (i) the amalgamation of the Ministry of Education (MOE), the Ministry of University Affairs (MOUA) and the Office of National Education Commission into a single Ministry of Education with a new administrative structure; (ii) the right to education for all citizens; (iii) the freedom to provide educational services; (iv) the recognition of formal, nonformal and casual education; and (v) the quality assurance requirements (National Education Act B.E.2542, 1999).

The newly founded Higher Education Commission (HEC) was instructed to supervise Thailand's higher education system. More autonomy was given to universities, institutes and colleges, which were to assume legal entity governed by their own boards. Freedom in academic matters and management was granted. The two major challenges faced by higher educational institutions being "efficient governance and institutional management" (Kiranandana et al., 1999).

A new Office of National Educational Standards and Quality Assessment (ONESQA), an autonomous agency with its own board of directors, was created receiving funding from the government's budget. An Office for Education Reform (OER) was established in 1999 with the duty to oversee this aforementioned transition period (Office for Education Reform, 2002).

The massification of higher education in Thailand continues; after a period of preparation, 46 public institutes which were originally teacher-training and technical colleges, were upgraded to become full-fledged universities in 2004. A separate Commission for Vocational Education (CVC) was established to oversee the "below-degree" phenomenon of postsecondary education which would, however, eventually expand.

There has also been an expansion of graduate education toward master and doctoral degrees. There were only 1,434 students at the graduate levels in 1994; while in 2001 there were 43,238 master degree students and 2,441 doctoral degree candidates. Research activities in the universities have correspondingly increased.

In addition, there is much more diversification of courses offered. Many courses for working students, in particular teachers and government's employees, have been devised. Some are at the diploma and master degree levels. Master courses in business administration, management, and information technology have become very popular.

In Thailand the issue of the quality of higher education has been of great concern for many years. At major universities, attempts have been made to change from didactic teaching to more effective forms such as learning by "inquiry," "self-learning," "experience-based learning," "problem-based learning," and "research-based courses." Teacher-training and faculty development programs have been activated in many places. The Ministry of University Affairs (MUA) provided faculty development grants and from 1990 to 1998, 13,706 grants totaling THB3,801 million were allocated (Ministry of University Affairs, 1998).

Before 1999, the Ministry of University Affairs screened and approved the opening of new courses in public universities. In conformity with requirements, adequate resources and preparedness were considered; no audit or follow-up assessment was in place. For private institutions, initial screening and periodic evaluation were required by law. Actual practices were to say the least, not effective and these requirements, nevertheless, put up barriers against their further development. For professional education such as medicine, engineering, law, and others, professional councils served to assure quality standards. In the past decade, several universities have voluntarily embarked upon "quality control" and "assurance measures."

With the creation of the Office of National Educational Standards and Quality Assessment (ONESQA) according to the new law, the framework for quality of higher education institutions is set with 8 standards and 28 indicators, with each institution doing self-assessment. A team of surveyors, consisting of senior educators appointed by the Office, visits the institutions for periodic assessment and accreditation. However, it is still at its very early stages of development, and experience must be gained for it to become really effective (Office of National Educational Standards and Quality Assessment, 2002).

The rapid and massive expansion of the higher education system has put great strain on limited human and financial resources, thus quality will undoubtedly be affected. For instance, some new universities have only 100 faculty members to oversee several thousand full- and part-time students. Laboratory facilities and communication connectivity are also limited, and the human resource management system would be an added big barrier. In spite of the institutional autonomy for financial and personnel management, the education system is still a long way from being a "functional system" that ensures quality (Suwanwela, 1996, 2002a).

From 2002 to 2006, Thailand is stipulated in the 9th Economic and Social Development Plan, which emphasizes the human element as a basis for a society of quality, learning, wisdom, and harmony. National competitiveness and self-reliance with strong community and efficient management are the main aims. Higher education is seen as an important enabling factor for the success of the National Development Plan. Relevance of higher education is being examined with regard to the national development plan. Structural and functional as well as budgetary reforms are taking this into consideration (Ministry of University Affairs, 2001).

In order to assure quality of institutions in the higher education system, there have been attempts to diversify institutions into "research universities," "teaching universities" and "community-oriented institutions." However, emphasis is put on research as an important educational tool and as a means for spreading "location-specific knowledge." All different varieties of university are encouraged to promote research development.

Information and communication technology (ICT) is recognized as a new "tool of opportunity" for handling problems facing higher education. The Open universities, Ramkamhaeng and Sukhothai Thammathiraj universities, have extensively used television in their educational programs. Interactive classes are however limited. It has been recognized for some time that computer literacy and competency would be an essential tool for future university graduates; connectivity and the use of it in education has been addressed, but appears to be lagging behind. From the outset universities created their own ports; the Intranet system and computerized classrooms are in use in many places. In 1994, the Inter University-Network (UniNet) was established. It has been providing wide bandwidth linkage among higher educational institutions in Thailand as well as abroad. It was later linked with the National Education Network (EdNet), which covered basic education schools. In 1996 universities were given approbation to build information technology campuses in 31 provinces (where no university existed) but due to the economic crisis of 1997, the project has been delayed. Television conferencing has, however, been established and is being used by some universities.

In 2002, the government approved the establishment of the National E-learning Centre and promoted its use. The Continuing Education Centre (CEC) of the Chulalongkorn University and a number of other establishments have recently embarked on the e-learning mode.

With the event of budgetary reform and university autonomy the rules have been modified as follows concerning the allocation of the government's budget to higher education institutions:

- *An operating* budget is meant for the provision of education is to be provided based upon per head cost.
- *Through submission* of successful proposals research funds could be allocated by various ministries and agencies.
- *Investment funds* must be in line with the government's policies and priorities.
- *Higher education institutions* must depend more on other sources of income.
- *Students' fees* at public institutions would be increased, while scholarships for needy and bright students must be provided in order to prevent the downfall of equity in higher education.
- *Intellectual property management*, donation, and efficiency improvement must be sought after.

Research funding in the universities needs to be diversified and the government's budgeting system changed. Tuition fees and educational support to students should stem from scholarship and loans to students and the development budget of universities should be submitted and considered on merit. Research funding should be sought from granting agencies such as the National Research Council, the Thailand Research Promotion Fund, the Science and Technology Development Agency as well as contracts from various ministries.

As from 2003, a new Comprehensive Research Support Scheme has also been introduced to provide for large visionary research and development ventures according to national priorities. Some research support should come out of the provincial budget so that research can serve provincial development plans and activities. Nongovernmental sources of research support from both inside and outside the country must be sought. Discussion is also ongoing for the establishment of a central national knowledge planning mechanism with the support of several sector-mechanisms (Thai National Health Foundation, 2003).

Globalization and Free Trade in Higher Education

Thailand has committed ten of its sectors, including education, in the General Agreement on Trade in Services (GATS) multilateral schedule. In addition Thailand is a part of the Association of Southeast Asian Nations (ASEAN) Asian Free Trade Agreement (AFTA), and is negotiating a bilateral Free Trade Agreement (FTA) with Australia, China, India, Japan, and the United States of America. A certain number of countries namely Australia, Japan, and the United States of America have expressed interest in Thailand's higher education sector.

Even prior to these agreements Free Trade in Higher Education had been present for many years but recently has greatly expanded. Many joint programs between Thai and foreign institutions have provided courses toward degrees at all levels, as well as nondegree courses. A number of universities abroad have set up branch campuses in Thailand, where parts of, or all of, the programs are carried out. Business administration, management, computer science, and technology as well as engineering are the more attractive topics in great demand. Franchises in language and in computer training have started operating. Examination programs for certification in English language competency as well as in basic education are a profit-making business. So far, there are, however, legal and procedural barriers in the way. Strict scrutiny and a rigid frame of reference as applied under the Private Higher Education Act have deterred many.

With bilateral Free Trade Agreements in particular the situation is changing. It appears that the Thai Party for Negotiation based mainly in the Ministry of Commerce and the Ministry of Foreign Affairs is not well prepared regarding higher education. The bargaining power of Thailand is also limited. Besides, there is inadequate expertise upholding the regulatory measures that would be required.

Since the market mechanism for higher education is an imperfect one, negative impacts arising from free trade in higher education can be serious. With less competitive capability of local providers, they might face difficulties to the point of closing down. International providers also bring in global culture, which would be beneficial for some, but it can pose as a threat for the traditional culture and value system.

Thailand is not just a receiving country, but can serve as an advantageous partner when dealing with developing neighboring countries. Extending the benefit of higher education in Thailand to neighboring countries would serve to enhance the good relationships between countries. The Thai Department of Technical and Economic Co-operation (DTEC) has, in the past decade, given scholarships to students from these countries to enable them to come and study in Thailand.

"Research" and the "generation of knowledge" can be important areas for cross-national activities and cooperation. Development of research capability is very much needed in the region. Knowledge base for mutual understanding—and regional as well as global harmony must be sought. Collaborative research in the region and exchange of researchers and students would be a positive move to make concerning the current trend in cross-border higher education.

Conclusion

Thailand is at present undergoing massive reform of its higher education and research systems. New structures and functions are in place and many more are being established and negotiated. Changes are required covering many dimensions in order to cope with the massification which is still going on, as well as new challenges being faced.

Due to the economic crisis and its slow recovery, financial resources are limited. Information and communication technology (ICT) poses both a "threat" and an "opportunity." Free trade in higher education appears to be inevitable, and the questions are "How to change threat into opportunity? "and" How to minimize unwanted consequences?" The most crucial factor would be the human element where, however, mentalities have to be modified. The development of "management capability" is also a crucial necessity.

Thorough research into the universities themselves is necessary to find better and final solutions during this transitory period and which could, indeed, provide new knowledge that would create an opportunity for equitable and sustainable development.

References and Works Consulted

Areegul, S. (2000) "Evaluation Report of the Research System in Thailand, 2000." Bangkok, National Research Council of Thailand Publication.
Institute of Health Services Research Annual Reports (1994) Bangkok, Institute of Health System Research, Ministry of Public Health.
—— (2000) Bangkok, Institute of Health System Research, Ministry of Public Health.
—— (2003) Bangkok, Institute of Health System Research, Ministry of Public Health.
Kiranandana, T. et al. (1999) "Research Report on the Preparedness for University Autonomy." Bangkok: Office of the Permanent Secretary, Ministry of University Affairs.
Ministry of University Affairs (MUA) (1991) "Annual Report." Bangkok: Ministry of University Affairs.
—— (1998) "Three Decades Report." Bangkok: Ministry of University Affairs.
—— (2001) "9th Higher Education Development Plan." Bangkok: Ministry of University Affairs.
National Education Act B.E.2542 (1999) Bangkok, Ministry of Education.
National Science and Technology Development Agency (1998) "Annual Report 1998." Bangkok: National Science and Technology Development Agency, Ministry of Science and Technology.
Office for Education Reform (OER) (2002) "Report on the Progress of Education Reform." Bangkok: Office of Education Reform, Ministry of Education.
Office of the National Education Commission (ONEC) (2000) "Capability of Thai Education in the Global Setting." Bangkok: Office of the National Education Commission, Ministry of Education.
ONESQA (Office of National Educational Standards and Quality Assessment) (2002) "Framework for External Quality Evaluation for Higher Education." Bangkok: Office of National Educational Standards and Quality Assessment, Ministry of Education.
Suwanwela, C. (1996) "Re-Engineering Higher Education." Case study of the Chulalongkorn University, Bangkok: Chulalongkorn University Press.

Suwanwela, C. (2002a) "Crisis of Public Universities in Thailand." In the monograph "The Crisis of Higher Education in Thailand and the Ways to Overcome Problems." Bangkok: Office of the Permanent Secretary, Ministry of University Affairs.
—— (2002b) *Thai Higher Education: A Synthesis*. Bangkok: Chulalongkorn University Press.
—— (2002c) *World Research System and Thai Research System*. Bangkok: Thai National Health Foundation (NHF) Publication.
Thai National Health Foundation (2003) "Report of the Study on Thai Research System." Bangkok: Thai National Health Foundation (NHF) Publication.
Thailand Research Fund (TRF) (2000) "Annual Report." Bangkok: Thailand Research Fund Publication.

CHAPTER TWELVE

CONCLUSION

RESEARCH MANAGEMENT IN THE POSTINDUSTRIAL ERA:
TRENDS AND ISSUES FOR FURTHER INVESTIGATION

V. Lynn Meek

Introduction

In the Asia Pacific Region, as elsewhere in the world, research and scientific researchers are engaged in long-term processes of responding to "the impacts of the heightened political profile of science, linked to more questioning public attitudes to science, and to the expansion of science and higher education" (Morris, 2004:2). Just after World War II, Vannevar Bush, vice-president and dean of MIT and scientific advisor to the then president of the United States of America articulated the rationale for sustained and substantial public support of basic research in both universities and research laboratories outside the academy. Initially, many Western nations established complex research systems, devoted mainly, though not exclusively, to advancement of knowledge and scientific discovery for its own sake. The linear notion of investment automatically leading to scientific discovery, knowledge transfer, and innovation came into question in the latter part of the twentieth century and support for pure, nonutilitarian research shifted more toward funding of applied knowledge production having demonstrable economic impact (Rip and Van der Meulen, 1996). This has further heightened the importance of science and higher education as one of its primary institutionalized promoters.

While many nations of the Asia Pacific Region may not have experienced the science policy development cycle of Western countries, the link between knowledge production and social and material welfare is being firmly established. The case studies presented in this book clearly demonstrate the growth in higher education and research in a number of Asia Pacific countries. As in the West, in most countries, various intermediatory bodies have been established between government and the research enterprise to channel both funding and policy. Every system examined in this book has experienced rapid expansion of higher education participation over the last decade, fueled at least in part by the awareness of the need for high-level skills required for participation in the evolving "knowledge-based economies." In all of the countries, research management and knowledge transfer have become important issues at both the sector and institutional levels.

This concluding chapter will not summarize the developments in higher education and research that have been presented in the various country studies. This task has been admirably completed by the case study authors themselves. Rather, this book will conclude with a general discussion of some of the broad trends and issues shaping higher education, research, and knowledge in the Asia Pacific Region. It is important to have an understanding of the key global and international pressures the systems more or less face in common, even if their responses are diverse. The purpose of this task is twofold. *First*, there are some generally held assumptions about the universality of the changing nature of higher education and research that deserve questioning. *Second*, there is a need to suggest a research agenda for the Asia Pacific Region that identifies important knowledge gaps in how the higher education systems of the region are responding to change.

The discussion in this chapter focuses on five topics: (i) commodification of knowledge and rise of the "knowledge-based economy"; (ii) globalization and internationalization of higher education; (iii) managerialism and the marketization of higher education; (iv) appropriate levels of public financial support; and (v) diversification of higher education functions in a "knowledge-based society." The conclusion speculates that, while the commodification of knowledge and the development of the "knowledge economy and society" are transforming higher education, the modern university is a resilient institution and likely to survive for sometime in a recognizable form.

Commodification of Knowledge and Rise of the "Knowledge-Based Economy"

The increasing recognition of the importance of research and the training of a highly skilled workforce in positioning nations in a global "knowledge-based economy" at once elevates the importance of higher education institutions and threatens many of their traditional values. The process is part and parcel of the advent of the "postindustrial society" and the commodification of knowledge—commodification taken here to mean "the phenomenon in which nonmaterial activities are being traded for money" (Lubbers, 2001). Neave (2002:3) explains,

> Knowledge has always been power as well as a public good. Access to it and its role in innovation determine both the place of nations in the world order and of individuals in society. But, commodification displaces the creation and passing on of knowledge from the social sphere to the sphere of production. Displacing and reinterpreting knowledge under these conditions raise fundamental questions for the university above all, in the area of academic freedom and in the "ownership" of knowledge. They also pose questions about the ethical obligation to make knowledge freely available to those who seek it.

In the mid-1980s, Lyotard, 1984 (cited in Roberts, 1998:1) hypothesized that "the status of knowledge is altered as societies enter what is known as the *post-industrial age* and cultures enter what is know as the *post-modern age*." According to Roberts, knowledge "is becoming 'exteriorised' from knowers. The old notion that knowledge and pedagogy are inextricably linked has been replaced by a new view of knowledge as a *commodity*." Or as Oliveira (2002:1) puts it, "[T]here is an essential difference between 'science as a search for truth' and 'science as a search for a response to economic and political interests!' " Lyotard (1984) again (cited in Roberts 1998:1–2) states:

> Knowledge is and will be produced in order to be sold, it is and will be consumed in order to be valorised in a new production: in both cases, the goal is exchange. Knowledge ceases to be an end in itself, it loses its "use-value.". . . Knowledge in the form of an informational commodity indispensable to productive power is already, and will continue to be, a major—perhaps the major—stake in the worldwide competition for power.

According to the OECD (1996:3), "[K]nowledge is now recognized as the driver of productivity and economic growth, leading to a new focus on the role of information, technology and learning in economic performance. The term 'knowledge-based economy' stems from this fuller recognition of the place of knowledge and technology in modern . . . economies." Several writers have extended the concept, arguing that science and research are transforming the whole of the social structure, creating a knowledge-based society of global proportions. Concepts depicting this transformation are formulated by Gibbons and his colleagues (1994) in terms of Mode 1 and Mode 2: Science, and later Mode 2: Society (Nowotny et al., 2001). Etzkowitz and his colleagues provide the less ambitious conceptualization of the "triple helix," representing the complex interplay between universities, government, and industry in the innovation framework (Etzkowitz and Leydesdorff 2001). These concepts will be outlined later in the section entitled "Diversifications of Higher Education Functions in a Knowledge-Based Society."

There is clearly a reciprocal relationship between the massive and unprecedented expansion of higher education during the second half of the twentieth century and global economic restructuring based on the advent of "postindustrial" or "knowledge" society. In "postindustrial society," knowledge supersedes agriculture and manufacturing as the main means for wealth production and becomes the primary resource of society. It is not that agriculture and manufacturing disappear, but rather that technology has made both agriculture and manufacturing so efficient that they demand the attention of only a minority of the workforce (Perkin, 1991). However, it is wise to remember that "postindustrial," "knowledge-based society" is not a phenomenon that has suddenly been sprung upon the world with the advent of the new millennium.

The American sociologist Daniel Bell coined the term "post-industrial society" as far back as 1962, and predicted the replacement of factory workers by "knowledge workers" as the primary producers of wealth. Bell (1974:xi) presented the original formulation of the concept of the postindustrial society at a forum on technology and social change in Boston in 1962. At about the same time, Kerr (1982) in outlining "The uses of the university" argued powerfully that the exponential expansion of knowledge was opening the academy to the broader interests of society in an unprecedented fashion that would transform the university forever. Since these early speculations the knowledge economy has indeed become a global reality. And, on a global scale, wealth and prosperity have become more dependent on access to knowledge than access to natural resources.

As the "knowledge-based society" continues to develop, market relations based on knowledge production increasingly permeate all aspects and institutions of society, and the university is faced with a growing number of competitors in both research and training. Also, the commodification of knowledge is impacting heavily on the internal social structure of the scientific community. What is at question is the continuing importance and centrality of the university as knowledge is increasingly brought within market and political exchanges.

According to Scott (1997:14), former editor of the "Times Higher Education Supplement" and current vice-chancellor of Kingston University, "[U]niversities have been absorbed into, been taken over by, market relations." Or put another way, "higher education systems are no longer simply 'knowledge' institutions, reproducing the intellectual and human capital required by industrial society; they are becoming key instruments of the reflexivity which defines the post-industrial (and post-modern) condition" (Scott, 1995:117).

Scott indicates that the interesting sociological question is whether higher education institutions, universities in particular, will continue to be recognized as such in the twenty-first century. Commentators such as Scott and Gibbons see the university losing its monopoly over knowledge production to the extent that the institution may eventually disappear as an identifiable form. Scott, for example, provocatively entitles a 1998 journal article "The End of the European University" (Scott, 1998).

There can be no doubt that the different roles and functions ascribed to the university are becoming highly complex, and the academy will need to more effectively share some of its key functions with other institutions in society. Partners and competitors will be found amongst private sector R&D companies, corporate training departments, for-profit private education providers, and so on. But again there is nothing new about this. In a 1967 publication prophetically entitled "Toward the Year 2000," Bell, in arguing that the "major new institutions of the society will be primarily intellectual institutions," listed the research university as only one example of research and intellectual entities of various kinds and went on to state that "no single kind may dominate, though perhaps the universities may be the strongest because so many problems get thrown at them, and they are immediately available for the kinds of tasks that were not there before" (Bell, 1967:32). In another publication, Bell is more unequivocal: "[T]he university increasingly becomes the primary institution of the post-industrial society" (Bell, 1974:245–246).

The academy has never had a complete monopoly over the production or dissemination of knowledge and its "traditional" approach to science, as depicted by Mode 1 Knowledge production (Gibbons et al., 1994, see below) and is itself a relatively recent phenomenon (Rip, 2002). For example, historically in many countries some professions, such as medicine in the United Kingdom and engineering in Portugal, had to fight protracted battles before being let into the universities. Moreover, at least in terms of the funding of basic research—particularly outside bioscience—there is evidence to suggest that in the United States of America the research universities have actually increased their dominance over the last couple of decades (Geiger, 2000). In reference to the Gibbons typology, insofar as Mode 2: Knowledge production is less hierarchical and more democratic and socially distributed, one can recognize a parallel with "indigenous knowledge production" (see K. H. Thaman, chapter nine, in this volume) and thus argue that Mode 2 precedes Mode 1: Knowledge production rather than the other way around. And, while new technology is influencing the way we teach within universities, there is no evidence that the emerging "virtual university" will replace the physical campus and its traditional function of socialization of the next generation of social, political, and scientific leaders.

Nearly everywhere, the university is required to find a new legitimacy while retaining essential traditions. Where, in the past, universities had a sense of shared intellectual purpose (at least to a certain degree), bolstered by the security of centralized funding and control, at present they are confronted by a much more complex, fluid, and varied environment that articulates different, and sometimes conflicting demands, creating new realities. Consequently, new distributions of authority emerge, new accountability relationships arise, and a new dynamic within policy fields develops.

The higher education systems of the Asia Pacific Region analyzed in this volume clearly demonstrate that they too face many of the general trends and issues outlined above. As elsewhere, universities in the region are expected to more closely serve the development needs of the nations. The is as true for the more developed countries like Australia and Japan as for the developing nations of Indonesia and the Philippines. Contributing to the "knowledge economy" through research products and training a skilled labor force is a key responsibility of all of the higher education systems examined in this book. China and its massive expansion of higher education clearly demonstrate the great expectations governments have for the contribution of higher education to social and economic development.

Thus, there are many global trends and issues impacting the higher education systems of the Asia Pacific Region to which imperatively they must respond. This does not imply that each system or institution within a system responds to these global forces in the same way. In fact, the opposite appears to be the case. There is far more diversity in the way in which higher education systems and institutions respond to similar environmental forces than commonly presumed (Amaral, et al., 2003; Currie et al., 2003). The country case studies presented in this volume also demonstrate great diversity in the structure, character, and function of the higher education systems of the region. Nonetheless, more research is required on how the various systems are responding to the commodification of knowledge and how the "knowledge economy" is being structured in different locations. More data needs to be collected on who participates in the knowledge economy in different countries and the structural barriers to participation by certain groups, be they based on socioeconomic status, gender, or ethnicity. On the other hand, it is also important to have more information on what approaches to knowledge production and transfer work best in promoting the "knowledge-based economies" of the different countries. A comparative study of best practices in the region regarding ownership of intellectual property and knowledge transfer would be most worthwhile.

Globalization and Internationalization of Higher Education

The terms globalization and internationalization of higher education are not easily defined and the two terms are often confused with one another. According to Altbach (2002:1), "globalization refers to trends in higher education that have cross-national implications," such as student markets, internet-based technologies, the global knowledge economy, and massification of higher education, while internationalization "refers to the specific policies and initiatives of countries and individual academic institutions or systems to deal with global trends," such as international student recruitment. However, for most practical purposes, it is impossible to keep the two phenomena entirely separate conceptually.

Green (2002:1) maintains that "international higher education" is an "umbrella term for the various institutional programmes and activities that are international in nature, such as student and faculty exchange, study abroad, international development activities, foreign language studies, international studies, area studies, joint degree programmes and comparative studies, among others." Knight (1999) divides international higher education into four approaches: (i) the activity approach (involving discrete activities along the lines described by Green); (ii) the competency approach (which stresses "the development of skills, knowledge, attitudes and values"); (iii) the ethos approach (emphasizing "a campus culture that fosters internationalisation"); and (iv) the process approach ("the integration of an international dimension into teaching, research and service"). To this list, one could add (i) the business approach (which emphasizes the maximization of profit from international student fees); and (ii) the market approach (with its stress on competition, market domination, and deregulation). No one approach to international higher education dominates all the others. Even the market approach, which has been so strong for a number of years, is now being moderated by quality assurance concerns and a negative popular reaction to economic globalization.

As Meek reports in this volume (see chapter four), Australia is one country that has embraced globalization with open arms, particularly with respect to the recruitment of international fee-paying students and the considerable revenue for higher education that they generate. Global industrial networks and new transport and communication technologies both enable and require higher education institutions to operate globally, and it is essential that "universities are able to participate as effective, highly regarded players in the international education and research architecture that underpins the new global knowledge economy" (Go8, 2002:14).

The benefits of international higher education are considerable: a more aware and culturally sensitive citizenry; collaboration with colleagues overseas and building international research networks; generation of export income, and so on. But international higher education is not without its problems either. With the rapid increase of international higher education, both in Australia and elsewhere, questions of the maintenance of a desirable level of quality have come about. Also, clearly, some nations benefit much more than others from international higher education, fueling tensions between the richer and poorer countries in this respect. As summarized in the introduction to this volume, the divide between the wealthier and more powerful nations with respect to knowledge production and related aspects of globalization was a key concern of the participants in the Scientific Committee's 1st Regional Research Seminar for Asia and the Pacific, Tokyo, Japan, 2004. The country case studies also demonstrate that the island nations of the Pacific, Indonesia, and the Philippines also face pressing problems in this regard. Nonetheless, many of the nations of the Asia Pacific Region are rapidly building their research capacity and expanding their higher education systems. India, for example, is a world leader in aspects of ICTs, China is building world-class research universities, and Japan is continuing to expand various innovative research programs.

Until relatively recently, the dominating influence on international higher education in the second half of the twentieth century was aid to developing countries (Altbach, 2001). This may appear to have changed with the rise of the highly lucrative international student market. But, still, the flow of international students is mainly from least developed countries to more developed nations, and underlying many of the concerns over international higher education is the continuation of the domination of the wealthy Western countries of the least developed nations of the world (Altbach, 2002). Such concerns are motivating much of the disquiet over the actions of the World Trade Organization (WTO) and policies such as the General Agreement on Trade in Services (GATS). At the same time, some countries, such as India, are engaged in creative solutions to such perennial problems as "brain drain."

The structure of aid to higher education in developing countries has changed dramatically over the past 4 decades, although many of the problems inherent in the donor/recipient relationship linger. "It is problematic to separate trade from aid" (Phillips and Stahl, 2000:4). In an era of globalization, we see many donor agencies as having strong links to international corporations, such as the Ford Foundation. It is worth considering whether Carnegie, Ford, Rockefeller, and other such foundations may become the "watchdogs" of the higher education sector worldwide with their agendas and regulatory frameworks gaining dominance where there is weakening control of the nation-state. The aid relationship has been internationalized, creating its own distinct problems. For example, the World Bank (WB) and the International Monetary Fund (IMF) through their rules and regulations already determine much of the policy of higher education in many countries. There dominance, however, is also under challenge in some countries.

Higher education institutions have always encouraged international cooperation and the free flow of ideas and professional personnel between countries. They have appreciated that science and scholarship do not recognize national boundaries and that progress in research will be facilitated by effective international sharing of ideas and discoveries.

Curiosity still motivates a large number of students to seek study abroad but, in the latter half of the twentieth century, international higher education endeavors have increasingly become tied to the development of global markets and worldwide economic restructuring. The internationalization of higher education is expanding and, as the production of wealth increasingly becomes based on knowledge rather than mechanization, it can be expected that the exploitation of international "knowledge markets" will assume even greater importance. In fact, exclusion of the least developed countries from or subordination to international knowledge-markets will reinforce the so-called north-south divisions.

Neave (2002:1) argues that most experts analyze internationalization by "studying what is happening in the advanced economies—Northern America, Western Europe, Asia and Australasia. These societies are relatively stable in their political, social and institutional make up." For the developed societies internationalization and globalization generally, while singling dramatic change, also have the potential for bringing about substantial reward. Elsewhere, "where the nation state is less than half a century old, where rivalries—some ethnic, others about beliefs—are all too easily inflamed, the upheaval that stands in the offing would seem even more devastating. It threatens the stability needed to build well-performing systems of higher education" (Neave, 2002:1). Altbach, too recognizes the inequalities in the current trends in internationalization in higher education:

> A few countries dominate global scientific systems, the new technologies are owned primarily by multinational corporations or academic institutions in the major Western industrialized nations, and the domination of English creates advantages for the countries that use English as the medium of instruction and research. All this means that the developing countries find themselves dependent on the major academic superpowers. (Altbach, 2002:1)

Despite dramatic growth in student numbers, many commentators argue that the full potential of international higher education cooperation and the free flow of ideas is not being fully realized. More could be done to promote the free flow of scientific information and research findings, and to assist developing nations through fellowships and grants. The needs of the least developed countries—many of them small—are serious and the prospects for substantial change in these countries, at least in the short-term, are limited unless the more developed countries are able to increase their technical assistance and other aid. There is clearly the Matthew effect at play in international higher education relations: "Guidance comes from the Gospel according to St Matthew, more particularly from the Parable of the Talents . . . 'To him that hath, it shall be given. To him that hath not, it shall be taken away even that which he hath.' A true Revelation divinely inspired!"(Neave, 2002:3).

While the internationalization of higher education and the way in which it is being institutionalized are tied directly to expansion brought about by the development of "postindustrial society" and the advent of the "knowledge economy," this does not mean that there has not always been an international aspect to higher education. Nor does it mean that an international community is replacing local customs, cultural ties and traditions; the present nature of the world political order does not sustain such a conclusion. Nonetheless, competition amongst nations for the control and productive utilization of knowledge is increasing. Moreover, the power to shape and influence the direction of internationalization and cooperation in higher education clearly rests with the larger and more powerful institutions and systems of the advanced countries. These countries do not present a united front; they compete amongst themselves for foreign students, control of knowledge, and influence in the international higher education arena. The developing countries are not powerless in this relationship, but the balance is tipped toward the more advanced industrialized nations. That said the danger of lumping all developing countries into the same category needs to be avoided, particularly with respect to the Asia Pacific Region. Many of the nations of this region, such as China and India, are increasingly having a worldwide economic impact, and other countries, such as Malaysia, Singapore, and Thailand are rapidly changing their status vis-à-vis the industrialized Western nations.

It is often the case with broad social movements that for every action in one direction, there is an "equal and opposite reaction" in the other. With the collapse of the iron curtain and the former Soviet Union at the end of the 1980s, the victory of the free market was celebrated almost everywhere. With respect to education, Johnstone (1998:4) writes that "underlying the market orientation of tertiary education is the ascendancy, almost worldwide, of market capitalism and the principles of neo-liberal economics." But the beginning of the new millennium was witness to a growing, and sometimes violent, protest against the further ascendancy and globalization of the market. These protests, such as those in Seattle (United States of America), and later in Genoa (Italy), and Melbourne (Australia), are directed at such programs as the General Agreement on Trade in Services (GATS) in particular and the spread of the global market economy in general. The War in Iraq has added another unsettling dimension to international relations, sparking protests of various extremes. It is not the place here to argue the rights and wrongs of these protests, but to merely point to them as at least partial evidence that world domination by market capitalism and the principles of neoliberal economics may not be as inevitable as some have assumed.

It is interesting to note that even the former senior vice-president of the World Bank, Joseph Stiglitz (in Rutherford 2001:2) has argued that "the fact that knowledge is, in central ways, a public good and that there are important externalities means that exclusive or excessive reliance on the market may not result in economic efficiency." The debates concerning the public and private good of higher education and knowledge production have been prominent in all of the higher education systems of the Asia Pacific Region examined in this book. Nonetheless, there is a need for more research on the public good benefits of higher education and on the relative balance in return on investment in higher education in relation to other education sectors. In recent years, the neoliberal market approach to higher education has influenced much of the thinking on the purposes and structures of the sector. Weaknesses in some of the more ideological extremes of the neoliberal market approach require analysis. This is particularly the case with respect to higher education management and coordination structures in a number of countries. But before turning to this topic in more detail in the section entitled "Managerialism and the Marketization of Higher Education" another important area for further research with respect to globalization and internationalization deserves mentioning—that is, cultural modifications and absorption of international trends and forces.

As Cummings demonstrates in Chapter two of this book, the countries of the Asia Pacific Region have histories and strong philosophical traditions substantially different from those of the Western nations. The modern university is a Western institution which so far appears to flourish in the Eastern climates where it has been transplanted. But not enough is known about the compatibility of the idea of the Western university with Eastern culture and philosophy or its impact on "indigenous knowledge systems" (see K. H. Thaman, chapter nine in this volume).

Managerialism and the Marketization of Higher Education

As part of a wider agenda of public sector reform, new approaches to higher education steering, and coordination have replaced government control with that of the market. Many of these changes have been characterized by a move away from tight state control and regulation of higher education toward a less restrictive state supervisory model (Neave and van Vught, 1991). Governments in promoting a climate of deregulation and decentralization have introduced policies that stress the necessity of strong management, competition, user-pays, and performance-based funding. The trend toward marketization and privatization of public sector higher education has been well established over the past decade or more, and is clearly visible in the language of policy documents (students as customers and clients, knowledge as a product or commodity, price and quality relations, etc.) and in their implementation: the introduction of tuition fees, performance-based funding, and conditional contracting. The introduction of market-like mechanisms changes the relationships among the actors in higher education and makes the environment in which universities must operate all the more fluid and turbulent.

Santiago et al. (2005) argue that a primary characteristic of the recent marketization of higher education in many nations has been the introduction of new management practices. Fueled by the realities of financial stringency on the one hand, and government demands for greater efficiency, effectiveness, and accountability on the other, higher education institution managers are increasingly relying upon a corporate style of management, drawing principles and practices from the "new public management" movement, which in turn is having a profound impact on the structure and character of the academic profession.

It appears that the modern university has shifted its orientation from social knowledge to market knowledge and that the "development of a market-oriented university supersedes academic decision making" (Buchbinder, 1993:335). According to Newson (1993:298), "These new forms of decision making fundamentally undermine a conception of the university as an autonomous, self-directing, peer-review and professional-authority based institution, and thus changes the politics of how academic work is accomplished." There is a view that in responding to market opportunities in a highly competitive global economic environment "traditional governance often works against making decisions fast enough to capitalise on new opportunities and avoid threats" (Green, et al., 2002:9).

The reorientation of higher education management, according to Currie and Newson (1998), is a product of the impact of globalization on higher education where market ideology and market or quasi-market modes of regulation are fused with a set of management practices drawn from the corporate sector. It is within this context that traditional forms of academic management are seen as obsolete and inefficient, progressively being replaced by practices based on the criteria of economic rationality. Articulation of clear and affirmative missions, commercial marketing, strategic management, and strategic and financial planning are today seen as fundamental management instruments for competing successfully in the higher education "industry" (Santiago et al., 2005). Much of this competition is directly related to the management of researchers and the commercialization of their products.

The notion of research management is a relatively recent phenomenon for most higher education systems. Traditionally university research was controlled by the researchers themselves, performed mainly within a discipline-based structure and was purported, if not actually, to be very much at the pure basic end of the research spectrum. However, certainly over the past decade or so, the relevance of university research has been questioned, and pressure brought to bear to make it more economically and socially relevant. The fact that much of the research income is performance-dependent, coupled with government pressures for commercialization of the intellectual property produced within the academy, has led to the centralization and professionalization of research management in many countries. The exploitation of knowledge has become a central concern for higher education managers.

Some of the higher education literature tends to assume that "traditional" collegial approaches to academic management are being replaced everywhere by corporate-orientated management processes. The empirical evidence, however, suggests a much more complex and diverse picture. Not only in higher education but in the general public policy literature, the notion that new public management will continue to colonize all in its path is being questioned, along with the adequacy of many of the movement's basic tenets (Meek, 2003:8–10). Moreover, some of the management technics borrowed from the private sector have not lived up to expectation. Dearlove (2002), for example, maintains that strategic planning is stronger at the level of intentions than at the level of implementation, for the latter entails the substantial involvement of the professoriate. Birnbaum (2000) claims that failure of various management technics imported into higher education from the corporate sector may be due to their normative, political character that ignores the traditional characteristics of higher education institutions.

Currie et al. (2003:187) conclude that management structures appear to be path-dependent and that there is no reason why all higher education institutions will respond the same way when exposed to managerialism. They also note that differences between formal managerial rules and rhetoric and day-to-day practices are often quite pronounced, requiring an in-depth look at what is happening within individual institutions. As Currie et al. (2003) argue, "There is a tendency to strengthen executive leadership or to centralize certain aspects of decision making, but this has not automatically changed the role of academics in decision making." Another important variable in how higher education institutions respond to managerialism identified by Currie et al. is tradition. Not only is there strong pressure within institutions to "maintain previously established procedures," but also the robustness of tradition needs to be understood within the broader cultural and political context. Also with respect to tradition, Altbach et al. (2001) caution against too easily relinquishing academic traditions—academic freedom, commitment to open enquiry in teaching and research, pursuit of knowledge for its own sake, and so on—upon which much of the past success of higher education has been built.

Mouwen (2000:55) writes that "the entry of the 'market' into the academic world is inevitable and will, whether we like it or not, lead to fundamental changes in the strategic position of universities in the landscape of higher education." But how fundamental and wide-ranging the changes are, first of all, *are empirical questions*, to which Gumport (2003:2) adds the following:

- What happens to management culture, resource allocation and traditional academic governance when markets increasingly influence institutional decision making?
- What is the impact of market forces on academically important fields that do not have a lucrative proximate market?

- Under what conditions do market forces work against vis-à-vis an institution's commitment to building a diverse faculty or student body?
- What happens to legislative influence when state revenue constitutes a declining share of public institutions' revenues?

These research questions are as fundamental to the higher education systems of the Asia Pacific Region as anywhere else in the world.

The idea of a comprehensive and one-directional influence of "managerialism" on higher education should be received with some skepticism. For on the one hand, as Winter and Sarros (2001:19) maintain, "[G]aining the support of the 'managed' will not be an easy task when many academics feel personally threatened by the tenets and practices of managerialism." On the other hand, we must be careful to neither exaggerate the degree to which the managerial has usurped the collegial nor the extent of contestation this has caused without reference to specific practices under peculiar cultural and political circumstances (Santiago et al., 2005). Of course, much of managerialism is grounded in the global (Currie et al., 2003), and the ultimate task, as Marginson and Rhoades (2002:305) point out, is to understand how organizational and human agencies "operate simultaneously in the three domains or planes of existence—global, national, local—amid multiple and reciprocal flows of activity."

As elsewhere, higher education institutions in the Asia Pacific Region are being asked to be better managed, transparent, accountable, and efficient in delivering educational and research services in relation to the support they receive. Australia is a key example of government demanding of higher education institutions that that they do more for less. But the higher education institutions in the other countries, too, are at once being given more autonomy to pursue their own respective distinctive goals and missions, while being held more directly accountable for public expenditure, with funding based on performance appraisal in one form or another. The corporatization of universities in Japan and the institutional autonomy movement in Indonesia are examples of management reforms in higher education based on the dual notions of autonomy and accountability. These and the higher education systems in the other countries of the region are experiencing pressures for stronger management in order to more effectively compete in an increasingly competitive global market.

But much more research that takes into account differences as well as similarities amongst the institutions and systems of the Asia Pacific Region on the consequences of changes to the management of higher education and the introduction of market relationships is required. Also, such research needs to be more empirically based rather than lead by normative assumptions on the universality of managerialism and marketization. Market coordination of higher education is not necessarily deleterious per se. The problem rests not so much with the introduction of such measures as strong management, competition, user pays, budget diversification, and entrepreneurial incentive, as with how these policies are actually constructed and implemented in specific contexts. In the absence of sustained research on various aspects of higher education management, governance, and coordination, including the management of research, there is the danger that many of the higher education systems of the Asia Pacific Region will merely repeat the mistakes made worldwide.

Appropriate Levels of Public Financial Support

As indicated throughout this chapter, higher education has an important and special role to play in the "knowledge-based economy" and "postindustrial society." "While the relationships between knowledge creation, innovation and economic growth are complex, there is widespread acceptance that the creation, distribution and exploitation of knowledge can lead to jobs growth and better standards of living" (DEST, 2003:115–116). The size and level of national participation in higher education is determined by, amongst other things, the rate of return on investment in higher education, and, "internationally, views on this issue have been strongly influenced by the emergence of the global knowledge-based economy" (Phillips et al., 2002:21). At the Australian 2001 Innovation Summit it was stated that

> [we] are in the midst of a revolution from which a New Order is emerging. The solutions of past decades will not suffice in the new knowledge age. Intangible assets—our human and intellectual capacity—are outstripping traditional assets—land, labour and capital—as the drivers of growth. If we are to take the high road, a road of high growth based on the value of our intellectual capital, we need to stimulate, nurture and reward creativity and entrepreneurship. (Backing Australia's Ability, 2001)

In a recent policy analysis, the OECD (2002) reports on the relative importance of human capital and education in economic growth. The report states that

> [t]he accumulation of physical capital and human capital is important for economic growth, and differences between countries in this respect help significantly to explain the observed differences in growth patterns. In particular, the evidence suggests that investment in education may have beneficial external effects that make social returns to schooling greater than private returns. (OECD, 2002:135)

And, further on, the report maintains that "the improvement in human capital seems to be a common factor behind growth in recent decades in all OECD countries." and states that

> [t]he magnitude of the impact on growth found in this analysis suggests that the social returns to investment in education may be larger than those experienced by individuals. This possibly reflects spill-over effects, such as links between levels of education and advances in technology, and more effective use of natural and physical resources, and implies that incentives for individuals to engage in education may be usefully enhanced by policy to reap maximum benefits for society as a whole. (OECD, 2002:135–137)

Another OECD publication (1998:7) reports that

> [i]nvestment in human capital is at the heart of strategies in OECD countries to promote economic prosperity, fuller employment, and social cohesion. Individuals, organizations and nations increasingly recognize that high levels of knowledge, skills and competence are essential to their future security and success.

Phillips et al. (2002:22–23) observe that

> [a] decline in public investment in tertiary education could perhaps be justified if there was evidence of negative social rates-of-return, but this is not the case.... Tertiary graduates, including higher education graduates, remain in demand in the labour market, and despite concerns about possible credentialism, graduates still enjoy substantially higher rates of earnings than the general population [approximately 50 percent higher in 1999 (OECD, 2002:123, Table A13.1)]. Overall, the OECD reports that both the private and social internal rates-of-return to tertiary education are "generally well above the risk-free real interest rate," i.e. tertiary education represents a good investment for both individuals and society as a whole.

Thus, it can be argued that "substantial and sustained growth in participation and investment levels in tertiary education would be sound economic and social policy" for all nations (Philips et al., 2002:23). The full potential of a higher education system to contribute to both the social and economic welfare of a nation cannot be achieved through governments abrogating their "duty" to adequately support it. As noted above, a common trend internationally is for sources, other than from government, to assume a greater proportion of overall higher education funding. But much more research is necessary to determine the optimal balance between public and private support of higher education as well as the optimal size of particular systems. Some countries in the Asia Pacific Region, it seems, are leaving future growth of higher education to the private sector, with little regard for the long-term social and economic consequences. There is an urgent need in several countries of the Asia Pacific Region for debates, informed by research and empirical evidence, on the amount of financial support of higher education that should come from the public wealth. Possibly for far too long the assumption that higher education primarily benefits the individual and therefore should be paid for by the consumer has gone unchallenged.

Diversification of Higher Education Functions In a "Knowledge-Based Society"

As mentioned in the first Section, Gibbons and his colleagues observe the transformation of the modern university in terms of the transition from Mode 1 to Mode 2: Knowledge production. Mode 1 is traditional science, hierarchical, strongly disciplinary-based, and elitist. Mode 2 is much more mass-oriented, democratic, and dispersed, characterized by "weakly institutionalized, transient and hierarchical organizational forms" (Johnston, 1998), and by

> fluidity, changing research teams, distributed research more generally; discovery in the context of application and transdisciplinarity; ... irrelevance of traditional disciplines; new forms of quality control ... ; contested expertise and (social) robustness as the new ideal; and the needed recontextualization (in society) of science and the institutions of science. (Rip, 2002:46)

The thesis of Mode 2: Knowledge production has been criticized as "simplistic in its projections," establishing a somewhat false dichotomy (Rip, 2002:45); ignoring the importance of past applied research (Godin, 1998); and ignoring "analyses that have highlighted the variety of functions played by disciplinary knowledge and discipline-based research training" (Henkel, 2002:59). Henkel notes that Rip (2000) argues that "the concept of Mode 1 can be seen as a 'lock-in' that exaggerates the rigidity of boundaries in which academic research practices are pursued and so threatens the heterogeneity required for advancement of knowledge that can tackle social as well as scientific problems of the future." Moreover, according to Henkel (2002:59):

> While it may be true that in some fields such as biological sciences interdisciplinary collaboration has proliferated and is an important driver of innovation, it has even there been concentrated within a relatively limited framework. It is most obviously explained by reference to the striking changes in these sciences themselves, triggered by the discovery of DNA and all that has followed from that. Again, such developments and the increased policy emphasis on inter-disciplinarity have not prevented many academics from continuing to see their discipline as having a critical role in their normative and epistemic identities. (Henkel, 2002:59; see also Grigg et al., 2003)

Nonetheless, as Rip (2002:46) states, while one can have doubts about the overall Mode 2: Knowledge production thesis, "many of the changes that are identified are important enough to take into account."

The proponents of Mode 2: Science and society draw an interesting parallel between Mode 1: Knowledge production and the elite university and Mode 2: Knowledge production and the mass university. "If true, this has important implications for the university. Its social and scientific roles, instead of being in tension (whether between the stasis of the elite university and the dynamics of progressive science, or between the open engagement of a democratic higher education and the disengagement of 'disinterested' science), may also be starting to overlap" (Nowotny et al., 2001:82). They argue that the scientific and social roles of the university, rather than being mutually exclusive, are actually mutually sustaining under a Mode 2: Environment:

> The development of higher education and research policies in many countries has been based on the belief that it is necessary to insulate the scientific functions of the university from its social functions, often equating the former with "elite" and the latter with "mass" education. The intention often has been to create a clearer separation between research, in which the elite university still plays an important but no longer exclusive role, and the higher education . . . of mass student populations where such a separation either does not exist, or to reinforce it, where it does exist, by encouraging the emergence of more differentiated systems. (Nowotny, et al., 2001:84–85)

The proponents of Mode 2 argue against institutional differentiation, particularly that based on past conceptualizations of "academic" or "vocational" and "scientific" or "professional." Nowotny et al. (2001:87–88) argue that "high-profile attempts to maintain, or promote, differentiation between research-led and access-oriented institutions have not always been successful because of the political difficulties such attempts create." It is difficult to segregate research-led universities from access-oriented higher education institutions in open, democratic societies, which may "help explain the tendency to seize on quasi-market, or actual market, solutions."

As a consequence, "not only has the number of 'researchers' within higher education systems increased as a result of the expansion of these systems since 1960; research is now undertaken in a wider range of non-university settings which extend far beyond free-standing research institutes or dedicated R&D departments into government, business, community and the media" (Nowotny et al., 2001:88).

Clearly, "the old division of labour between fundamental and applied or problem-oriented research has almost disappeared, and with it, the functional distinctions between universities, public labs and industrial and other private research" (Rip, 2002:46). Also, according to Rip (2002:47), "The contrast between fundamental (and scientifically excellent) research . . . and relevant research . . . is not a principled contrast. It has more to do with the institutional division of labour, than with the nature of scientific research."

Moreover, there can be little dispute that many societies have become more knowledgeable and that with the advent of the World Wide Web and other forms of modern telecommunications, access to knowledge has become more widespread and nearly instantaneous. At the same time, society has successfully challenged the elite position, autonomy, and exclusivity of many professions, including academic researchers. The knowledge society is simultaneously more dependent upon science and less trustful of it and its proponents—"enhanced understanding [of science] tends to diminish rather than increase public confidence" (Henkel, 2002:60; see Wynne, 1995; Bauer et al., 1997).

Nonetheless, differentiation both within and between institutions remains an important policy question and, contrary to the Mode 2 thesis, the empirical evidence strongly suggests that research remains the primary differentiator. One of the most important areas for research in the Asia Pacific Region, as elsewhere, is how best to differentiate higher education systems to serve the multiplicity of needs and demands of mass higher education and the "knowledge-based economy."

As argued in the introduction to this book, no country can afford to fund all of its universities as "world-class research universities," and institutional emulation often results in second-rate imitations. Moreover, those institutions that emulate research universities without sufficient resources to adequately do so cannot provide their students, particularly their research students, with appropriate tuition. Emulation of research universities also diverts institutions away from engaging in extensive programmatic diversity that appears imperative for mass higher education (Meek, 2000). In many countries, the numbers and quality debate about higher education has led to the conclusion that "quality can be protected by creating a hierarchy of institutions catering to different sections of the market" (*The Economist*, 1997). The important research question is how to foster diversity by preventing institutions from converging on a single preconceived "gold standard" of what is proper higher education.

Nowotny et al. argue that:

> Under Mode-2 conditions, the distinction between research and teaching tends to break down. This happens not only because the definition of who now qualifies as a research actor must be extended far beyond the primary producers or research, but also because the reflexivity of Mode-2: Knowledge production transforms relatively closed communities of scientists into open communities of "knowledgeable" people. (Nowotny et al., 2001:89–90)

Much of the argument plays on the meanings of "research" and "knowledge." In adopting a fairly traditional definition of research (publications, grants, patents, etc.), then questions of differentiation of function both within and between institutions remain important concerns. Arguments based on the fundamental importance of the nexus between teaching and research in higher education are often self-serving, particularly when we take into account that in all higher education systems something like 80 percent of the research output is produced by 20 percent of the staff. Since research attracts prestige, everyone wants a share, despite the legitimacy of their claim.

From a research management point of view, it does not appear that research is a democratic, widely dispersed activity, and, as stated above, one might question the nexus between teaching and research—at least in terms of research that generates external funding. A case probably can be made that all university staff should be engaged in scholarship at a high level, which means staying informed about the latest research in their areas of expertise. However, with respect to research itself, concentration and selectivity appear to be the order of the day.

This issue is not so much the separation of teaching and research. The evidence suggests that this occurs regardless. What is important is the policy context that structures the way in which the boundaries between teaching and research are created and maintained. It is probably true that "economic growth is affected not only by the quantum of funding but by the way funds are allocated (for example, in terms of the institutions, fields and industries to which they are directed, and the mechanisms used to finance research) and by knowledge dissemination and research commercialisation practices that are adopted" (DEST, 2003:115). On the other hand, a narrow priority-driven and overly utilitarian approach to public support for research may in the long term be counterproductive. Henkel (2002:64) cites investigations that suggest that "since outcomes of inquiry are often wholly unpredictable, imposing limits in terms of future relevance or applicability is likely to reduce, rather than enhance, the social or economic benefits it may generate."

Contrary to the Mode 2 thesis, the research university is unlikely to disappear, though it is being transformed as it interacts with an increasingly complex and turbulent environment. According to Rip (2002:49), "[T]he key challenge is to diversify and recombine its components, both cognitively and institutionally, into what I call a post-modern university. Such a university will include overlaps and alliances with Centres (of excellence and relevance), public laboratories of various kinds (themselves on the move!) and various private organizations managing and performing research. The boundaries between the university and the outside world are porous, and such 'porosity' is sought explicitly."

While the boundaries between the university and the outside world may be becoming more porous, this does not necessarily mean the comprehensive dissolution of the normative structures that maintain scientific communities specifically and academic organizations generally. According to Henkel (2002:60), the extent of category collapse implied in such theses as Mode 2: Knowledge production is questionable, although "it is not necessary to subscribe wholesale to a post-modern perspective to perceive a variety of ways in which the boundaries between academic and other worlds are being blurred and to conclude that this is a growing trend."

Even Mode 2 proponents recognize that the university, though under mass conditions of higher education, must remain relatively stable in order to continue to fulfil two primary functions: the production of the next generation of researchers and generator of cultural norms" (Nowotny et al., 2001:93).

Thus, the question of diversification *versus* homogenization of higher education institutions and systems is one of the most important areas for further research for all nations of the Asia Pacific Region. There is evidence to suggest that formally differentiated higher education systems produce more "true" institutional diversity than unitary systems (Meek et al., 1996). But this proposition has yet to be tested in the Asia Pacific Region outside of Australia and New Zealand (Codling and Meek, 2003). Research on diversity also should include degrees of concentration of research funding, the primacy of the teaching-research nexus in higher education, and the distribution of higher education institutions between the public and private sectors.

Conclusion

Bertelsen observes that "the commodification of higher education to serve the market is revolutionizing our entire practice, from institutional image through to management, jobs and curriculum," and goes on to state that

> once they have conceded that knowledge is a commodity to be traded, universities become subject . . . to the full and ruthless protocols of the market. Time-honoured principles of truth and intellectual rigour are rapidly superseded by cost-effectiveness and utility, and market rules are systematically applied. First, research is only done if it creates new products, and courses which do not feed job skills are a waste of time. So managers dutifully prioritise "core business" and eliminate "peripheral" activities, and funding becomes an investment decision based on short-term production goals. (Bertelson, 2002:1)

Management in many institutions strongly promotes those areas of the enterprise that appear to turn a profit, while shedding investment in less lucrative activities, such as the humanities, ancient and some modern languages, and so on. Given the decline of public funding and rising student numbers in a highly competitive and volatile market, institutional leaders may well indeed argue that they have no other choice.

In the past, academic loyalty was first and foremost to the discipline and to disciplinary norms concerning the definition and production of knowledge (Becher, 1989; Clark, 1983; Gouldner, 1958). With the commodification of knowledge, that loyalty has come under challenge from powerful groups both within and without the academy demanding loyalty first and foremost to the institution—that is, to the corporation that pays the bills (Meek, 2003). "Science policies, national and international have, in different degrees, been eroding academic autonomy since the early 1970s" (Henkel, 2002:58). Henkel goes on to state that the "landmark here is the Harvey Brooks Report for OECD (1971) which laid down the principles that governments rather than scientists must set over-riding research priorities and that the key driver of science policies must be the achievement of social and economic goals." In a similar vein, Slaughter and Leslie (1997:5) argue that

> [p]articipation in the market began to undercut the tacit contract between professors and society because the market put as much emphasis on the bottom line as on client welfare. The *raison d'etre* for special treatment for universities, the training ground of professionals, as well as for professional privilege, was undermined, increasing the likelihood that universities, in the future, will be treated more like other organizations and professionals more like other workers.

However, as argued above and as both Slaughter and Leslie (1997) and Henkel (2002) note, neither the academy in general nor the scientific community specifically have been passive participants of these changes. Probably science has transformed society more so than governments have transformed the university. Clearly, the commodification of knowledge has led to new types of relationships within the academy based on what Slaugher and Leslie (1997) refer to as "academic capitalism," and the academic capitalist professor has become a powerful position within many universities. According to Henkel (2002:60), "academic scientists and the institutions in which they work have become more or less willing actors in a range of markets and so in the commodification of scientific knowledge." She goes on to state that "capacity for profit making sits alongside intellectual reputation as high value currency in an increasingly competitive academic labour market." But this does not mean that the university is being transformed out of all recognition.

Many of the scenarios applied to the future of the university, where they are not out rightly speculative (such as the replacement of the traditional campus by the "virtual university" or the disappearance of the academy altogether) display a regrettable element of ungrounded exaggeration. What should be treated as empirical questions requiring rigorous examination, such as the replacement of Mode 1: Science with Mode 2, tends to remain at the level of normative assertions. There can be little doubt that "postindustrial society" and the "knowledge-based economy" will demand even greater diversity from higher education institutions and systems wherever they are located. Society will impose new roles, pressures, and demands on higher education while simultaneously expecting the preservation of key traditional functions (Neave, 2000). Higher education institutions in turn will help shape the very society that generates these new and traditional expectations.

The modern university, whether located in the Asia Pacific Region or elsewhere, may be a victim of its own success. However, the university over hundreds of years has proved to be quite a resilient social institution. One hopes for a heightened awareness, particularly amongst governments, of both the importance of understanding the changing role of higher education in society and of the critical contribution higher education makes to shape a nation's future, both economically and socially. This book will have achieved much if it enhances awareness of the key issues facing the higher education systems of the nations of the Asia Pacific Region and stimulates more research into their cause and effect.

Acknowledgment

This chapter is based partially on a paper presented by the author at the 2nd Meeting of the UNESCO Regional Research Scientific Committee for Asia and the Pacific, New Delhi, India, September 8–9, 2003.

References and Works Consulted

Altbach, P. G. (2001) "Why Higher Education Is Not a global commodity." *The Chronicle of Higher Education* (electronic version).

—— (2002) "Perspectives on Internationalisation in Higher Education." Resource Review. International Higher Education Series, The Boston College Center for International Higher Education, 27.

Altbach, P., P. Gumport, and D. B. Johnstone (2001) *In Defense of the American Public University*. Baltimore: Johns Hopkins University Press.

Amaral, A., V. L. Meek, and I. Larsen, eds. (2003) *The Higher Education Managerial Revolution?* Amsterdam: Kluwer Academic Publishers.

Backing Australia's Ability (2001) *Australian 2001 Innovation Summit*. Canberra: Commonwealth of Australia.

Barnett, R. (2000) *Realizing the University in An Age of Supercomplexity*. Buckingham: SRHE and Open University Press.

Bauer, M., J. Durant, and E. N. Gaskell (1997) "Europe Ambivalent on Biotechnology." *Nature*, 387:845–847.

Becher, T. (1989) *Academic Tribes and Territories*. Milton Keynes: Open University Press.

Bell, D., ed. (1967) *Toward the Year 2000*. Boston: Houghton Mifflin, 32.

—— (1974) *The Coming of Post-Industrial Society*. London: Heinemann, 5–246.

Bertelsen, E. (2002) "Degrees 'R' Us—The Marketisation of the University," http://web.uct.ac.za/org/aa/chomsk.htm#Eve Bertelsen.

Birnbaum, R. (2000) *Management Fads in Higher Education, Where They Come From, What They Do, Why They Fail*. San Francisco: Jossey-Bass.

Buchbinder, H. (1993) "The Market Oriented University and the Changing Role of Knowledge." *Higher Education*, 26(3):331–347.

Clark, B. R. (1983) *The Higher Education System: Academic Organization in Cross-National Perspective*. Berkeley: University of California Press.

Codling, A., and V. L. Meek (2003) "The Impact of the State on Institutional Differentiation in New Zealand," *Higher Education Policy and Management*, 15:83–98.

Currie, J., and J. Newson, eds. (1998) *Universities and Globalization: Critical Perspectives*. London: Sage.

Currie, J., R. De Angelis, H. de Boer, J. Huisman, and C. Lacotte (2003) *Globalizing Practices and University Responses*. Westport, CT: Praeger.

Dearlove, J. (2002) "A Continuing Role for Academics: The Governance of UK Universities in the Post-Dearing Era," *Higher Education Quarterly*, 56(3):257–275.

DEST (2003) "The National Report on Higher Education in Australia (2001)." Canberra, DEST.

Etzkowitz, H., and L. Leydesdorff (2001) *Universities and the Global Knowledge Economy*. London: Continuum.

Geiger, R. (2000) "University-Industry Research Relationships: Trends and Issues Drawn from Recent USA Experience." Paper presented at Experts' Meeting "Research Management at the Institutional Level," OECD/IMHE, Paris, June 8–9.

Gibbons, M., C. Limoges, H. Nowotny, S. Schwartzman, P. Scott, and M. Trow (1994) *The New Production of Knowledge: The Dynamics of Science and Research in Contemporary Societies*. London: Sage.

Godin, B. (1998) "Writing Performative History: The New Atlantis?" Review of Gibbons, M. et al., *The New Production of Knowledge: The Dynamics of Science and Research in Contemporary Societies. Social Studies of Science*, 28(3):465–483.

Gouldner, A. (1958) "Cosmopolitans and Locals: Toward an Analysis of Latent Social Roles." *Administrative Science Quarterly*, 2:281–306, 444–467.

Green, M. F. (2002) "Remarks for Panel Session." Paper presented at the International Conference on internationalization of higher education: policy and practice, International Association of Universities, Lyon, Rhone-Alpes, France, April 12–13.

——— P. Eckel, and A. Barblan (2002) *Brave New (and Smaller) World of Higher Education: A Transatlantic View*. Washington: ACEnet.

Grigg, L., R. Johnston, and N. Milsom (2003) *Emerging Issues for Cross-Disciplinary Research: Conceptual and Empirical Dimensions*. Canberra: Department of Education, Science and Training.

Go8 (Group of Eight) (2002) "Submission to Higher Education Review." 14. http://www.dest.gov.au/crossroads/submissions/pdf/181.pdf (accessed July 31, 2005).

Gumport, P. (2003) "Beyond Dead Reckoning: Research Priorities for Redirecting American Higher Education," *International Higher Education*, Winter:1–3.

Henkel, M. (2002) "Current Science Policies and their Implications for the Concept of Academic Identity." In *Proceedings, Science, Training and Career, Changing Modes of Knowledge Production and Labor Markets*. Enschede: CHEPS/University of Twente, 55–70.

Johnston, R. (1998) *The Changing Nature and Forms of Knowledge: A Review*. Canberra: Evaluation and Investigations Program (DEST), 1998.

Johnstone, D. B. (1998) "The Financing and Management of Higher Education: A Status Report on Worldwide Reforms." Paper presented at the UNESCO Conference on Higher Education, UNESCO, Paris, October 5–9.

Kerr, C. (1982) *The Uses of the University*, 3rd ed. Cambridge: Harvard University Press.

Knight, J. (1999) "Internationalization of Higher Education." In J. Knight and H. de Wit, eds., *Quality and Internationalization in Higher Education*. Paris: Organisation for Economic Co-operation and Development, 13–28.

Lubbers, R. (2001) "Definition: Commodification." www.http://globalize.kub.nl/.

Lyotard, J-F. (1984) *The Post-Modern Condition: A Report on Knowledge*. Minneapolis: University of Minnesota Press.

Marginson, S., and G. Rhoades (2002) "Beyond National States, Markets, and Systems of Higher Education: A Glonacal Agency Heuristic." *Higher Education*, 43:281–309.

Meek, V. L. (1996) *The Mockers and Mocked: Comparative Perspectives on Diversity, Differentiation and Convergence in Higher Education*, L. Goedegebuure, O. Kivinen, and R. Rinne, eds. Oxford: Pergamon Press.

——— (2000) "Diversity and Marketisation of Higher Education: Incompatible Concepts?" *Higher Education Policy*, 13(1):23–39.

——— (2003) "Governance and Management of Australian Higher Education: Enemies Within and Without." In A. Amaral, V. L.Meek, and I. Larsen, eds., *The Higher Education Managerial Revolution?* Amsterdam: Kluwer Academic Publishers, 149–171.

Morris, N. (2004) "Scientists Responding to Science Policy: A Multi-Level Analysis of the Situation of Life Scientists in the UK." Doctoral dissertation, Enschede, University of Twente, Centre for Higher Education Policy Studies.

Mouwen, K. (2000) "Strategy, Structure and Culture of the Hybrid University: Towards the University of the 21st Century." *Tertiary Education and Management*, 6:47–56.

Neave, G. (2000) *The Social Responsibilities of the Universities*. Oxford: Pergamon Press.

——— (2002) "Globalization: Threat, Opportunity or Both?" *International Association of Universities Newsletter*, 8(1) (2002):1–3.

Neave, G., and F. van Vught, eds. (1991) *Prometheus Bound*. Oxford: Pergamon Press.

Newson, J. (1993) "Constructing the 'Post-Industrial' University: Institutional Budgeting and University Corporate Linkages." In P. G. Altbach and B. D. Johnstone eds., *The Funding of*

Higher Education: International Perspectives. Garland Studies in Higher Education, vol. Q. New York: Garland Publishing, 285–304.

Nowotny, H., P. Scott, and M. Gibbons (2001) *Re-Thinking Science: Knowledge and the Public in an Age of Uncertainty*. Malden, MA: Blackwell.

OECD (Organisation for Economic Co-operation and Development) (1971) "Science, Growth and Society: A New Perspective." Report of the Secretary-General's *Ad Hoc* Group on New concepts of science policy, chaired by Harvey Brooks, Organisation for Economic Co-operation and Development, Paris.

—— (1996) *The Knowledge-Based Economy*. Paris: Organisation for Economic Co-operation and Development.

—— (1998) *Human Capital Investment: An International Comparison*. Paris: Organisation for Economic Co-operation and Development.

—— (2002) *Education at a glance 2002*. Paris: Organisation for Economic Co-operation and Development.

Oliveira, L. (2002) *Commodification of Science and Paradoxes in Universities*. ISCTE, Lisbon: University of Lisbon, http://rektorat.unizg.hr/rk/Docs/uisa.html (accessed July 31, 2005).

Perkin, H. (1991) "History of Universities." In *International Higher Education*, an Encyclopaedia. New York: Garland.

Phillips, D., L. Cooper, Ch. Eccles, D. Lampard, G. Noblett, and P. Wade (2002) "Independent Study of the Higher Education Review." Sage 1 Report. Byron Bay, NSW: Phillips Curran, http://www.kpac.biz (accessed July 31, 2005).

Phillips, M. W., and A. Stahl (2000) "Internationalization of Education, Training and Professional Services: Trends and Issues." Working Paper, July 2000.

Rip, A. (2000) "Fashions, Lock-Ins and the Heterogeneity of Knowledge Production." In M. Jacob and T. Hellström, eds., *The Future of Knowledge Production in the Academy*. Buckingham: Open University Press.

—— (2002) "Strategic Research, Post-Modern Universities and Research Training." In *Proceedings, Science, Training and Career, Changing Modes of Knowledge Production and Labor Markets*. Enschede: CHEPS/University of Twente, 45–54.

Rip, A., and B. J. R. Van der Meulen (1996) "The Post–Modern Research System." *Science and Public Policy*, 23:343–352.

Roberts, P. (1998) "Rereading Lyotard: Knowledge, Commodification and Higher Education." *Electronic Journal of Sociology*, 3(3), http://www.sociology.org/content/vol003.003/roberts.html (accessed July 31, 2005).

Rutherford, J. (2001) "Let the Corporate Class Commence." *The Times Higher Education Supplement* (electronic version).

Santiago, R., T. Carvalho, A. Amaral, and V. L. Meek (2005) "Changes in the Middle Management of Higher Education Institutions: The Case of Portugal." *Higher Education*, 52(2):215–250.

Scott, P. (1995) *The Meanings of Mass Higher Education*. Milton Keynes: Open University Press.

—— (1997) "The Changing Role of the University in the Production of New Knowledge." *Tertiary Education and Management*, 3(1).

—— (1998) "The End of the European University." *European Review*, 6(4) (1998):441–457.

Slaughter, S., and L. Leslie (1997) *Academic Capitalism: Politics, Policies and the Entrepreneurial University*. Baltimore: Johns Hopkins University Press.

The Economist (1997) "Inside the Knowledge Factory." October 2, 1997. htpp://www.economist.com (accessed July 31, 2005).

Winter, R., and J. Sarros (2001) "Corporate reforms to Australian universities: views from the academic heartland." Clayton, Monash University (unpublished paper).

Wynne, B. (1995) "Public Understanding of Science." In S. Jasanoff, G. Markle J. Petersen, and T. Pince, eds., *Handbook of Science and Technology Studies*. London: Sage.

Index

accountability, 49, 53, 66, 72, 81, 108, 112, 114, 120–1, 123–4, 128–9, 136, 138, 144, 149, 156, 217, 222, 224
Agrawal, A., 177, 183
Alcala, A., 189, 199
Altbach, P., 12–13, 25, 28, 41, 125, 126, 128, 132, 133, 134, 155, 156, 168, 171, 217, 219, 220, 223, 232, 233
Amano, I., 155, 172
Amaral, A., 133, 217, 232, 233, 234
Anandkrishnan, M., 122, 123, 132
Apple, M., 113, 122, 127, 132
applied research, 4, 16, 18, 34–5, 44–5, 53, 55, 86–7, 99–100, 118, 122, 146, 227–8
applied science
 see applied research
Arciga, A. J. P., 186, 193, 194, 195, 200
Areegul, S., 206, 211
Arimoto, A., 20, 23, 153, 155, 157, 159, 160, 166, 168, 169, 170, 172
Asian Development Bank (ADB), 47, 62, 186, 187, 198
Asia-Pacific Economic Cooperation (APEC), 31
Association of Southeast Asian Nations (ASEAN), 31, 210
Australia,
 Australian Research Council (ARC), 50, 51, 54, 58, 59, 62, 75, 77, 78, 79, 83
 Australian Vice-Chancellors' Committee (AVCC), 68, 70, 71, 74, 80, 85, 86, 87, 88
autonomy, 11, 19, 29, 36–7, 50, 80, 82, 90, 109–110, 112, 120, 136–9, 144, 149, 156, 163, 188, 203–9, 211, 224, 228, 230

Backing Australia's Ability (2001), 54, 63, 76, 79, 80, 82, 88, 89, 225, 232

Bakalevu, S., 176, 183
Barnett, R., 65, 88, 232
Bartholomew, J. R., 27, 41
Basaruddin, C., 13, 24
basic research, 2–4, 16–7, 34–6, 44–5, 48, 52–3, 55, 59, 86–8, 96, 99–100, 104, 136, 141, 143–4, 150, 162, 205–6, 213, 216, 222
basic science
 see basic research
Bauer, M., 228, 232
Becher, T., 165, 172, 230, 232
Bell, D., 215, 216, 232
Ben-David, J., 154, 169, 172
Benson, C., 177, 178, 184
Bernardo, A. B. I., 14, 15, 24, 25, 188, 194, 197, 199
Bertelsen, E., 230, 232
Bhattacharya, S., 122, 132
Bhushan, S., 108, 132
Birnbaum, R., 223, 232
Blumenthal, D., 56, 62
Boyer, E. L., 170, 172, 197, 198, 199
Bozeman, B., 57, 62
brain circulation, 9, 155
brain drain, 9, 38, 49, 107, 116, 157, 168, 219
brain gain, 157
Brennan, J., 112, 132
Brink, J.R., 44, 62
Brokensha, W. D. M., 177, 183
Bualat, A. S., 186, 194, 200
Buchanan, J. M., 108, 132
Buchbinder, H., 222, 232
Bush, V., 34, 42, 213

Calvert, J., 55, 62
Campbell, I., 132
Carvalho, T., 234
centres of excellence, 37, 137, 161

Chanana, K., 18, 19, 23, 107, 119, 127, 128, 133
Chaudhuri, P. P., 115, 116, 133
Chen, X., 119, 133
Choi, J., 59, 62
Christie, A. F., 59, 62
Chubb, I., 73, 85, 88
Clark, B. R., 65, 88, 159, 160, 164, 168, 169, 172, 230, 232
Codling, A., 230, 232
colonialism, 28, 179
commercialization, 4, 16–7, 43, 45, 48, 56–8, 60–2, 76, 79, 86, 89, 101, 222–3
Commission on Higher Education (CHED), 21, 185–199, 200
commodification, 128, 214, 216–7, 230–1, 233, 234
competition, 8, 10, 12–13, 20, 42, 54, 61, 72–3, 78, 96, 99, 101, 108–9, 111, 114, 116, 122, 139, 143–4, 146, 150–1, 153, 156–7, 161, 169, 171, 204, 215, 218, 220, 222, 224
see also market
concentration and selectivity, 17–18, 72–3, 85, 88, 229
Cooper, L., 234
Cooperative Research Centre (CRC), 55, 71, 79
Council of Scientific and Industrial Research (CSIR), 19, 118
Couturier, L., 89
Crossman, P., 178, 183
Cultural Revolution, 96
Cummings, W. K., 27, 42, 116
Currie, J., 217, 222, 223, 224, 232

Dahlman, C., 45, 48, 62
Dawkins, J., 72, 89
De Angelis, R., 232
de Boer, H., 232
De la Cruz, A. P., 200
De la Rosa, R. V., 186, 188, 199
Dearlove, J., 223, 232
Delors, J., 177, 183
Department of Education, Science and Training (DEST), 50, 62, 67, 68, 69, 70, 75, 82, 83, 84, 85, 89, 225, 229, 232, 233
Department of Industry, Science and Resources (DISR), 78, 79, 89
Deregulation, 72, 188, 218, 222
see also market

developing countries, 1–3, 11, 14–16, 22, 98, 111, 219–220
Devisch, R., 178, 183
Devletoglou, N. E., 108, 132
Dewes, W., 176, 183
Dickson, S., 55, 62
Directorate General of Higher Education (DGHE), 135, 139, 140, 146, 148, 151
diversity, 7–8, 15, 49, 66, 72–3, 77, 80, 110, 134, 143, 148–150, 217, 228, 230–1
DLSU-Manila CHED Zonal Research Centre, 186, 194, 195, 199
Durant, J., 232
Dutta, S., 116, 133

Eccles, Ch., 234
Economic Commission for Asia and the Far East (ESCAFE), 135, 139, 140, 146, 148
Ehara, T., 155, 168, 172
entrepreneurship, 58, 63, 146, 150, 225
equity, 4, 15, 20, 30, 81, 109, 133, 137–9, 148–51, 198, 204, 209
Esteva, G., 180, 183
Etzkowitz, H., 55, 62, 215, 232
Eustace, R., 109, 133

Famularcano, P. A. C., 200
Farnham, D., 108, 109, 110, 112, 113, 128, 132, 133
Fatnowna, S., 179, 183
Fedrowitz, J., 132
Fell, C., 82, 89
Forum on Higher Education, Knowledge and Research, 1, 133, 198
see also UNESCO
free trade, 210–11
Fuess, S., 38, 42
funding
see market

Gaita, K. L., 62
Gallagher, M., 68, 78, 89
Garrett, R., 112, 133
Gaskell, G. N., 232
Geiger, R., 44, 58, 62, 63, 160, 164, 172, 216, 232
Gender, 4, 20, 133, 138, 141, 144, 217
see also equity
General Agreement on Trade in Services (GATS), 5, 25, 156, 198, 210, 219, 221

General Office of the State Council, 103, 104
Gibbons, M., 3, 20, 25, 49, 62, 63, 159, 160, 172, 215, 216, 226, 232, 233, 234
Gibbs, P., 112, 133
Gittleman, M., 62
Glasick, C. E., 198, 199
global
  see globalization
Global Education Digest, 46, 62
Global Scientific Committee, 3
  see also UNESCO
globalization, 2, 6–7, 11, 12, 15–16, 20, 28, 42, 45, 61, 66, 90, 107, 109, 111, 115, 121, 134, 153, 155–7, 166, 171–2, 179, 181–2, 184, 193, 199, 210, 214–15, 217–19, 220, 221–4
  see also internationalization
Godin, B., 227, 233
Goldfarb, B., 57, 62
Goldsworthy, A., 76, 89
Gouldner, A., 230, 233
graduate education, 20, 38, 112, 128, 136, 140, 144–5, 148–9, 151, 168, 170–1, 197–9, 205–7
Green, M. F., 218, 222, 233
Green Paper on Higher Education in China, 105
Grigg, L., 227, 233
Group of Eight (Go8), 218, 233
Gumport, P., 156, 172, 223, 232, 233

Haden, C. R., 44, 62
Hall, B. H., 55, 63
Harman, G. S., 14, 16–17, 24, 25, 43, 46, 51, 56, 63, 200
Hayden, M., 66, 89
Health Systems Research Institute (HSRI), 22
Henkel, M., 227, 228, 229, 230, 231, 233
Henrekson, M., 57, 58, 62, 63
Hicks, F. B., 42
Higher Education Contribution Scheme (HECS), 72–4, 75, 77, 79, 81
Hoppers, C. A. O., 180, 183, 184
Howard, J., 54, 63, 76, 80, 89
Howlett, M. J., 62
Huber, M. T., 132, 199
Huisman, J., 232

Indian Council of Social Science Research (ICSSR), 4, 5, 23, 118

indigenous knowledge, 7, 20–1, 175–183, 216, 221
Indiresan, P. V., 123, 133
information communication technology (ICT), 3, 9–10, 52, 61, 84, 111, 116, 193, 208, 211
innovation, 9–10, 27–8, 36, 43–5, 47–8, 50, 54–5, 61, 68–9, 74, 76–9, 80, 82, 84–8, 100, 101, 109, 138, 141, 178, 190, 205, 213–15, 227
Institutional Grants Scheme (IGS), 78, 85
intellectual property (IP), 4, 9, 19, 48, 55, 57–60, 79, 112, 140, 147, 178, 184, 209, 217, 223
International Monetary Fund (IMF), 219
internationalization, 7, 11, 108, 116, 134, 154, 214, 217–221

Jackson, J., 66, 89
Jiang Zhe Ming, 95, 104
Johnson, C., 30, 42
Johnston, R., 226, 233
Johnstone, D. B., 221, 232, 233

Kelso, R., 107, 133
Kemp, D., 76, 77, 78, 79, 89
Kenney, M., 55, 63
Kerr, C., 215, 233
Kimura, S., 42
Kiranandana, T., 201, 211
Knight, J., 218, 233
knowledge transfer, 4, 21, 52, 181, 202, 213–14, 217
knowledge-based economy/society, 1–2, 7, 9, 13, 20, 98, 137, 155, 158–60, 165–6, 170–71, 193, 213–17, 225–6, 228, 231
Kodama, F., 27, 42
Kogan, M., 110, 133
Kolawole, O. D., 176, 184
Koswara, J., 19, 24, 135, 140, 142, 143, 147, 151

labor force
  see labor market
labor market, 7, 45, 47, 49, 70, 109–110, 187, 192
Lacotte, C., 232
Lampard, D., 234
Lanvin, B., 133
Larsen, I., 232, 233
Larson, J., 181, 183
Lauden, J., 110, 133

Lauden, K., 110, 133
Lawton, D., 177, 183
Leggett, C., 107, 133
Leslie, L., 55, 63, 231, 234
Leydesdorff, L., 215, 232
lifelong education, 157–8
Limoges, C., 25, 232
Low, M., 42
Lubbers, R., 214, 233
Lyotard, J-F., 214, 215, 233, 234

Mack, G., 42
Maeroff, G. I., 199
managerialism, 214, 223–4
  see also new public management
Marginson, S., 28, 42, 224, 233
Market, 2, 4, 8, 13, 39, 45, 49, 50, 65–6, 74, 80, 85, 101, 103, 107, 110, 112–14, 122, 127–30, 147, 155–6, 161, 164, 166, 171, 210, 216, 218–19, 221–4, 227, 228, 230–1
mass higher education, 7–8, 22, 48, 100, 110–11, 121–2, 125, 127, 134, 155, 157–8, 172–3, 204, 207, 211, 218, 226–8, 230
massification
  see mass higher education
Meek, V. L., 1, 17, 18, 22, 23, 24, 65, 69, 73, 75, 79, 89, 90, 122, 133, 213, 218, 223, 228, 230, 232, 233, 234
Meiji, 27, 32, 157, 163
Mel, M., 176, 183
Mercer, D., 76, 89
Merton, R. K., 160, 167, 172
Milsom, N., 233
Ministry of Education
  (MOE–Indonesia), 141
Ministry of Education (MOE-Japan), 154
Ministry of Education (MOE-New Zealand), 51, 52, 63
Ministry of Education (MOE-PRC), 18, 91, 92, 93, 94, 96, 97, 98, 100, 101, 104
Ministry of Education
  (MOE-Thailand), 207
Ministry of Education, Culture, Sports, Science and Technology (MEXT), 11, 23, 158, 161, 162, 172, 173
Ministry of National Education (MONE), 135, 151
Ministry of Science and Technology (MOST–Indonesia), 135
Ministry of Science and Technology (MOST-PRC), 95, 101, 102, 105

mode 1 / 2 knowledge production, 20, 49, 159–60, 215–16, 226–31
  see also Gibbons
moonlighting, 20, 148
Morris, N., 213, 233
Mortimer, D., 76, 89
Motohashi, K., 59, 63
Mouwen, K., 223, 233
Mshana, R., 178, 184
multidisciplinary research, 49, 99, 191–2, 194

Nabobo, U., 176, 184
Nagano, H., 11, 23, 24
Nakayama, S., 28, 42
National Centre for Education and Research, 105
National Education Act B.E.2542, 207, 211
National Health and Medical Research Council (NHMRC), 50, 59, 79
National Institute of Advanced Industrial Science and Technology (AIST), 11
National Institutes of Health (NIH), 34
National Institute of Population and Social Security Research, 158, 172
National Research Council
  (NRC-Indonesia), 144
National Research Council (NRC-Thailand), 22, 204, 205, 209, 211
National Science and Technology Development Agency (NSTDA), 22, 205, 206, 211
National Science Board (NSB), 30, 32, 33, 36, 38, 42
National Science Foundation (NSF), 34, 168, 173
Neave, G., 65, 89, 214, 219, 220, 222, 231, 233
Needham, J., 28, 42
neoliberal, 221
  see also market
new public management (NPM), 22, 53, 222–3
Newman, F., 86, 89
Newson, J., 22, 232, 233
Nigavekar, A., 5, 23, 25
Noblett, G., 234
Nonaka, I., 111, 133
Norris, D. M., 11, 133
Nowotny, H., 25, 43, 63, 215, 227, 228, 230, 232, 234
Ntuli, P. P., 177, 184

Office for Education Reform (OER), 207, 211
Office of National Educational Standards and Quality Assessment (ONESQ), 207, 208, 211
Office of the National Education Commission (ONEC), 201, 203, 211
Oliveira, L., 215, 234
Onaka, I., 156, 173
Ordoñez, V. M., 188, 199
Organisation for Economic Co-operation and Development (OECD), 3, 16, 32, 33, 39, 44, 45, 47, 48, 49, 50, 51, 54, 58, 59, 60, 61, 63, 67, 68, 69, 70, 73, 74, 84, 85, 94, 173, 215, 225, 226, 230, 232, 234
Osaki, H., 161, 173

Padua, R. N., 187, 199
Panchamukhi, V. R., 5, 23
Patel, P., 55, 62
Paua, F., 133
peer review, 3, 8, 37, 51–2, 109, 222
Perkin, H., 109, 133, 215, 234
Peters, L., 55, 62
Phillips, D., 74, 75, 89, 225, 226, 234
Phillips, M. W., 219, 234
Pickett, H., 179, 183
Pierson, W., 169, 173
Pijano, C. V., 187, 199
Porter, M. E., 137, 151
Postiglione, G., 42
postindustrial society, 213–15, 220, 225, 231
priority setting, 11, 14, 16, 43, 49, 52–4, 61–2, 83, 86
private education, 9, 12–13, 18–19, 21–2, 27, 35, 46, 59, 67–8, 80, 84, 107–8, 112–4, 120–31, 136–9, 142, 146, 149, 155, 186, 189, 197, 199, 203–4, 208, 210, 216, 221, 223, 226
privatisation, 2, 4, 6, 12, 66, 81, 85, 107, 111, 113, 121, 128, 222
see also private education
Project 211, 92, 105
Project 985, 18, 94, 98
public good, 3, 18, 50, 88, 111, 214, 221

quality assurance, 8, 22, 138, 151, 156, 171, 187, 199, 200, 207, 218

Rajagopalan, S., 115, 133
Ralph, J., 76, 89
Ramsey, A., 89

ranking, 37–8, 42
Regional Scientific Committee for Asia and the Pacific, 1, 4, 5, 23, 133
see also UNESCO
Research Assessment Exercise (RAE), 8, 18, 51, 82
research culture, 13–15, 19–21, 25, 139–41
research infrastructure, 14, 21, 58, 78–9, 82, 84–5, 88, 148, 162
research management, 1, 9, 17, 19, 43, 53, 61, 73–4, 78, 135, 140, 148, 190, 196, 214, 222–3, 229
research policy, 1–4, 10, 15–17, 20, 23, 43–4, 61–3, 72, 75, 84–5
Research Training Scheme (RTS), 78, 85
research university, 8, 12, 13, 43, 45, 85, 97–8, 160, 164, 166, 168, 205, 216, 229
Rhoades, G., 224, 233
Rip, A., 213, 216, 226, 227, 228, 229, 234
Roberts, P., 214, 215, 234
Rosenberg, N., 57, 58, 63
Rowland, S., 128, 129, 133
Rutherford, J., 221, 234

Salazar-Clemeña, R., 21, 23, 24, 185
Sanga, K., 178, 184
Santiago, R., 222, 224, 234
Sarji, A., 32, 42
Sarmiento, J. E., 194, 199
Sarros, J., 224, 234
Scarry, J., 89
Schwartzman, S., 25, 232
Science and Technology Basic Plan, 161, 162, 167, 172, 173
science cities
see science park
science park, 36, 100, 105, 205
Scott, P., 25, 63, 65, 66, 89, 90, 128, 129, 134, 216, 232, 234
Senate Review, 76, 80, 82, 90
Shah, T., 132
Shanghai Jiao Tong University Institute of Higher Education (SJTUIHE), 42
see also ranking
Shinbori, M., 167, 173
Sizer, J., 57, 63
Slaughter, S., 55, 63, 231, 234
Smith, Adam, 108
Smith, L., 176, 178, 184
SnowChange.Org., 181, 184
Solow, R. M., 137, 151

Stahl, A., 219, 234
stakeholder, 17, 49, 89
Stallings, B., 29, 42
standards
    see quality
Stenberg, L., 47, 50, 59, 63
Stiglitz, J., 221
Stocker, J., 76, 89, 90
Suwanwela, C., 1, 22, 201, 204, 205, 208, 211, 212
Swedish International Development Co-operation Agency (Sida), 1

Tadjudin, M. K., 19, 23, 24, 135
Takeuchi, H., 111, 133
Tan, E. A., 188, 199
Taufe'ulungaki, A. M., 177, 178
Teaero, T., 176, 184
Teasdale, J., 177, 184
Teasdale, R., 177, 184
technological transfer, 12, 18
Teichler, U., 24, 185, 199
Thai National Health Foundation, 209, 212
Thailand Research Fund (TRF), 205, 206, 212
Thailand Research Promotion Fund (TRPF), 22, 209
Thakur, D. S., 110, 134
Thakur, K. S., 110, 134
Thaman, K. H., 20, 21, 23, 175, 176, 177, 178, 182, 183, 184, 216, 221
The Economist, 65, 90, 228, 234
The Hindu, 126, 134
The New Scientest, 134
The Sydney Morning Herald, 67, 89, 90
Thompson, K. W., 107, 134
Tilak, J. B. G., 24, 113, 114, 115, 134
Trow, M., 25, 109, 110, 134, 173, 232

Umakoshi, T., 28, 41
UNESCO, 1, 3, 4, 5, 10, 23, 24, 25, 47, 62, 63, 89, 90, 95, 128, 133, 134, 177, 179, 180, 181, 183, 184, 188, 198, 200, 232, 233
United Nations University (UNU), 4, 10, 198
Universal Basic Education (UBE), 30
University Council (UC), 159, 162, 173

University Grants Commission (UGC), 5, 19, 23, 112, 134
University Grants Committee (UGC), 51, 53, 63
University of the South Pacific (USP), 175, 179, 180–2, 184
University Research Council (URC), 141
University Research for Graduate Education (URGE), 20
Ushiogi, M., 169, 173

Van der Meulen, B. J. R., 213, 234
Van Vught, F., 222, 233
Veblen, T., 109, 134
Vicencio, E. M., 186, 193, 194, 195, 200

Wade, P., 234
Wang Ling, 42
Webster, E. M., 62
Wei Yu, 18, 23, 24, 91, 104, 105
Werner, O., 177, 183
West, R., 76, 90
Whiston, T. G., 44, 63
Wills, P., 68, 76, 77, 90
Winter, R., 224, 234
Wirakusumah, S., 136, 151
Wood, F. Q., 73, 75, 76, 79, 80, 90
World Bank (WB), 29, 30, 42, 45, 48, 62, 94, 109, 111, 115, 134, 151, 181, 219, 221
World Conference on Higher Education (WCHE), 1, 185, 198, 200
World Trade Organization (WTO), 60, 156, 198, 219
World War I, 32
World War II, 29, 32, 34, 45, 154, 171, 172, 202, 213
Wynne, B., 228, 234

Yamamoto, S., 159, 170, 172
Yamanoi, A., 173
Yoshikawa, H., 11, 24
Yoshioka, H., 42
Yun, M., 59, 63

Zhao Xin-Ping, 97, 105
Zhou Ji, 94, 104, 105